Robert Kolker

HIDDEN VALLEY ROAD

Robert Kolker is the *New York Times* bestselling author of *Lost Girls*, named one of the *New York Times*'s 100 Notable Books and one of *Slate*'s best nonfiction books of the quarter century. His journalism has appeared in *New York* magazine, *The New York Times Magazine*, *Bloomberg Businessweek*, *Wired*, *O, the Oprah Magazine*, and The Marshall Project. He is a National Magazine Award finalist and a recipient of the 2011 Harry Frank Guggenheim Award for Excellence in Criminal Justice Reporting from the John Jay College of Criminal Justice in New York.

www.robertkolker.com

Robert Kolker is available for select
speaking engagements. To inquire about a possible
speaking appearance, please contact
Penguin Random House Speakers Bureau at
speakers@penguinrandomhouse.com
or visit www.prhspeakers.com.

Also by Robert Kolker

Lost Girls

Hidden Valley Road

INSIDE THE MIND
OF AN AMERICAN FAMILY

Robert Kolker

ANCHOR BOOKS
A Division of Penguin Random House LLC
New York

For Judy and Jon

FIRST ANCHOR BOOKS EDITION, MARCH 2021

Copyright © 2020 by Robert Kolker

All rights reserved. Published in the United States by Anchor Books,
a division of Penguin Random House LLC, New York. Originally published
in hardcover in the United States by Doubleday, a division of
Penguin Random House LLC, New York in 2020.

Anchor Books and colophon are registered
trademarks of Penguin Random House LLC.

Photograph on page 104 courtesy of Robert Moorman. All other
photographs courtesy of Lindsay Galvin Rauch and Margaret Galvin Johnson.

The Library of Congress has cataloged the Doubleday edition as follows:
Name: Kolker, Robert, author.
Title: Hidden Valley Road : inside the mind of an
American family / Robert Kolker.
Description: First edition. | New York : Doubleday, 2020. |
Includes bibliographical references and index.
Identifiers: LCCN 2019028466
Subjects: LCSH: Galvin family. | Schizophrenics—United States—Biography. |
Schizophrenia—Genetic aspects. | Schizophrenia—Treatment—United States—
History. | Schizophrenics—Family relationships—United States. |
Mentally ill—Care—United States—History.
Classification: LCC RC514 .K648 2020 | DDC 616.89/80092—dc23
LC record available at https://lccn.loc.gov/2019028466

Anchor Books Trade Paperback ISBN: 978-0-525-56264-1
eBook ISBN: 978-0-385-54377-4

Author photograph © Jeff Zorabedian
Book design by Maria Carella

www.anchorbooks.com

Printed in the United States of America
10 9 8 7 6 5 4 3

The clearest way that you can show endurance
is by sticking with a family.

—ANNE TYLER

CONTENTS

A brother and sister walk out of their house together, through the patio door that opens out from the family kitchen and into their backyard. They're a strange pair. Donald Galvin is twenty-seven years old with deep-set eyes, his head shaved completely bald, his chin showing off the beginnings of a biblically scruffy beard. Mary Galvin is seven, half his height, with white-blond hair and a button nose.

The Galvin family lives in the Woodmen Valley, an expanse of forest and farmland nestled between the steep hills and sandstone mesas of central Colorado. Their yard smells of sweet pine, fresh and earthy. Near the patio, juncos and blue jays dart around a rock garden where the family's pet, a goshawk named Atholl, stands guard in a mews their father built years ago. With the little girl leading the way, the sister and brother pass by the mews and climb up a small hill, stepping over lichen-covered rocks they both know by heart.

There are ten children between Mary and Donald in age—twelve Galvin kids in all; enough, their father enjoys joking, for a football team. The others have found excuses to be as far from Donald as possible. Those not old enough to have moved away are playing hockey or soccer or baseball. Mary's sister, Margaret—the only other girl, and the sibling closest to Mary in age—might be with the Skarke girls next door, or down the road at the Shoptaughs'. But Mary, still in second grade, often has nowhere to go after school but home, and no one to look after her but Donald.

Everything about Donald confounds Mary, starting with his shaved head and continuing with what he likes most to wear: a reddish brown bedsheet, worn in the style of a monk. Sometimes he completes the outfit with a plastic bow and arrow that his little brothers once played with. In any weather, Donald walks the neighborhood dressed this way, mile after mile, all day and into the night—down their street, the unpaved Hidden Valley Road, past the convent and the dairy farm in the Woodmen Valley, along the shoulders and onto the median strips of highways. He often stops at the grounds of the United States Air Force Academy, where their father once worked, and where many people now pretend not to recognize him. And closer to home, Donald has stood sentry as children play in the yard of the local elementary school, announcing in his soft, almost Irish lilt that he is their new teacher. He only stops when the principal demands that he stay away. In those moments, Mary, a second-grader, is sorrier than ever that her world is so small that everyone knows that she is Donald's sister.

Mary's mother is well practiced at laughing off moments like these, behaving as if nothing is strange. To do anything else would be the same as admitting that she lacks any real control over the situation—that she cannot understand what is happening in her house, much less know how to stop it. Mary, in turn, has no choice but to not react at all to Donald. She notices how closely both her mother and father monitor all of their children now for warning signs: Peter with his rebellion, Brian and his drugs, Richard getting expelled, Jim picking fights, Michael checking out completely. To complain or cry or show any emotion at all, Mary knows, will send the message that something might be wrong with her, too.

And the fact is that the days when Mary sees Donald in that bedsheet are better than some of the other days. Sometimes after school, she comes home to find Donald in the middle of an undertaking only he can understand—like transplanting every last piece of furniture out of the house and into the backyard, or pouring salt into the aquarium and poisoning all the fish. Other times, he is in the bathroom, vomiting his medications: Stelazine and Thorazine and Haldol and Prolixin and Artane. Sometimes he is sitting in the middle of the living room quietly, completely naked. Sometimes the police are there, summoned by their mother, after hostilities have broken out between Donald and one or more of his brothers.

But most of the time, Donald is consumed by religious matters. Explaining that Saint Ignatius conferred upon him a degree in "spiritual exercise and theology," he spends much of every day and many nights reciting in full voice the Apostles' Creed and the Lord's Prayer and a list of his own devising that he calls Holy Order of Priests, the logic of which is known only to him. *D.O.M., Benedictine, Jesuit, Order of the Sacred Heart, Immaculate Conception, Mary, Immaculate Mary, Oblate Order of Priests, May Family, Black Friar, The Holy Ghost, Franciscan at the Convent, One Holy Universal, Apostolic, Trappist . . .*

For Mary, the prayers are like a faucet that won't stop dripping. "Stop it!" she shrieks, and yet Donald never does, pausing barely long enough to breathe. She sees what he's doing as a rebuke of her entire family, but mostly of their father, a faithful Catholic. Mary idolizes her father. So does every other Galvin child—even Donald did, before he got sick. When Mary sees her father coming and going from the house whenever he likes, she is envious. She thinks about the sense of control that her father must enjoy by working so hard all the time. Hard enough to get out.

It is the way her brother singles Mary out that she finds most unbearable—not because he is cruel but because he is kind, even tender. Her full name is Mary Christine, and so Donald has decided that she is Mary, the sacred virgin and mother of Christ. "I *am not!*" Mary cries, again and again. She believes that she is being teased. It would not be the first time that one of her brothers has tried to make a fool of her. But Donald is so unmistakably serious—so fervent, so reverential—it only makes Mary angrier. He has made Mary the exalted object of his prayers—bringing her into his world, which is the last place that she would ever want to be.

The idea that Mary comes up with, the solution to the problem of Donald, is a direct response to the rage she feels. Her inspiration comes from the sword-and-sandals epics that her mother sometimes watches on television. The idea starts with her saying, "Let's go up to the hill." Donald consents; anything for the sacred virgin. It continues with Mary suggesting that they build a swing on a tree branch. "Let's bring a rope," she says. Donald does as she says. And it concludes at the top of the hill, where Mary selects a tree, one of many tall pines, and tells Donald that she'd like to tie him to it. Donald says yes. And hands her the rope.

Even if Mary were to reveal her plan to Donald—to burn him at the stake, like the heretics in the movies—it is doubtful that he would react. He is too busy praying. He stands tightly against the tree trunk, lost in his own stream of words as Mary walks around the tree with the rope, circling and pulling until she believes he cannot break free. Donald does not resist.

She tells herself that no one will miss him when he's gone—and that no one will ever suspect her. She goes searching for kindling and brings back armfuls of twigs and branches, dropping them at his bare feet.

Donald is ready. If Mary really is who he insists she is, he can hardly say no. He is calm, patient, kind.

He adores her.

But on this day, Mary is serious only to a point. She has no matches, no way to make a fire. More crucially, she is not like her brother. She is grounded, her mind rooted in the real world. If nothing else, Mary is determined to prove that, not just to her mother, but to herself.

So she abandons her plan. She strands Donald on the hill. He stays up there, surrounded by flies and pasqueflowers, standing in place and praying for a very long time. Long enough that Mary gets some time to herself, but not so long that he doesn't come back down again.

SHE MANAGES A smile now when she thinks about it. "Margaret and I laugh," she says. "I'm not sure others would find it quite so funny."

On a crisp winter afternoon in 2017—forty-five years, a lifetime, after that day on the hill—the woman once known as Mary Galvin pulls her SUV into a parking space at Point of the Pines, an assisted living facility in Colorado Springs, and walks inside to see the brother she once fantasized about burning alive. She is in her fifties now, with the same bright eyes, though in her adulthood she has chosen to go by a different first name: Lindsay, a name she picked as soon as she left home—a determined young girl's attempt to make a break with the past and become someone new.

Lindsay lives a six-hour drive away, just outside Telluride, Colorado. She owns her own business, staging corporate events—working as hard as her father ever did, crisscrossing the state between home and Denver, where most of her events take place, and Colorado Springs, where she can tend to Donald and others in her family. Her husband, Rick, runs instructor training for the Telluride ski school, and they have two teenagers, one in high school and one in college. Anyone who meets Lindsay now usually doesn't see past her calm confidence, her easy smile. After years of practice, she has an artful way of pretending as if everything is completely normal, even when the case is quite the opposite. Only a tart, razor-sharp comment now and then suggests something else—something melancholic and immutable, simmering beneath the surface.

Donald is waiting for her in the first-floor lounge. Dressed casually in a wrinkled, untucked Oxford shirt and long cargo shorts, her oldest brother, in his seventies now, looks incongruously distinguished, with wisps of white hair at his temples, a cleft chin, and heavy black eyebrows. He could be cast in a gangster movie, if his voice weren't so gentle and his gait so stiff. "He has a little bit of that Thorazine shuffle still left, the way he walks," says Kriss Prado, a manager at the facility. Donald takes clozapine now, a sort of last-resort psychotropic drug with both a high rate of effectiveness and a high risk of extreme side effects—heart inflammation, low white blood cell count, even seizures. One of the consequences of surviving schizophrenia for fifty years is that sooner or later, the cure becomes as damaging as the disease.

When Donald spots his sister, he stands up, ready to leave. Usually, when Lindsay visits, it's to take him out to see other family. Smiling warmly, Lindsay says they're not going anywhere today—that she is there to see how he is doing and to talk with his doctors. Donald smiles, too, slightly, and sits back down. No one in his family comes to see him there but her.

Lindsay has had decades to make sense of her childhood, and in many ways that project continues. So far, she has learned that the key to understanding schizophrenia is that, despite a century of research, such a key remains elusive. There is a menu of symptoms, various ways the illness presents: hallucinations, delusions, voices, comalike stupors. There are specific tells, too, like the inability to grasp the most basic figures of speech. Psychiatrists speak of "loosening of associations" and "disorganized thinking." But it is hard for anyone to explain to Lindsay why, on a day like today, Donald is cheerful, even content, while on another day he is frustrated, demanding she drive him to the state mental hospital in Pueblo, where he has been admitted more than a dozen times over fifty years, and where he often says that he would like to live. She can only guess why, when Donald is brought to the supermarket, he always buys two bottles of All clothing detergent, announcing brightly, "This is the best body wash ever!" Or why, almost fifty years later, he still recites that religious litany: *Benedictine, Jesuit, Order of the Sacred Heart. . . .* Or why, for almost as long, Donald has consistently and unwaveringly maintained that he is, in fact, the offspring of an octopus.

The most dreadful thing, perhaps, about schizophrenia—and what most sets it apart from other brain conditions like autism or Alzheimer's, which tend to dilute and dissipate a person's most identifiable personality traits—is how baldly emotional it can be. The symptoms muffle nothing and amplify everything. They're deafening, overpowering for the subject and frightening for those who love them—impossible for anyone close to them to process intellectually. For a family, schizophrenia is, primarily, a felt experience, as if the foundation of the family is permanently tilted in the direction of the sick family member. Even if just one child has schizophrenia, everything about the internal logic of that family changes.

But the Galvins never were an ordinary family. In the years when Donald was the first, most conspicuous case, five other Galvin brothers were quietly breaking down.

There was Peter, the youngest boy and the family rebel, who was manic and violent, and who for years refused all help.

And Matthew, a talented ceramic artist, who, when he wasn't convinced that he was Paul McCartney, believed that his moods controlled the weather.

And Joseph, the most mild-mannered and poignantly self-aware of the sick boys, who heard voices, as real to him as life itself, from a different time and place.

And Jim, the maverick second son, who feuded viciously with

Donald and went on to victimize the most defenseless members of his family—most notably the girls, Mary and Margaret.

And, finally, Brian, perfect Brian, the family's rock star, who kept his deepest fears a secret from them all—and who, in one inscrutable flourish of violence, would change all of their lives forever.

THE DOZEN CHILDREN in the Galvin family perfectly spanned the baby boom. Donald was born in 1945, Mary in 1965. Their century was the American century. Their parents, Mimi and Don, were born just after the Great War, met during the Great Depression, married during World War II, and raised their children during the Cold War. In the best of times, Mimi and Don seemed to embody everything that was great and good about their generation: a sense of adventure, industriousness, responsibility, and optimism (anyone who has twelve children, the last several against the advice of doctors, is nothing if not an optimist). As their family grew, they witnessed entire cultural movements come and go. And then all the Galvins made their own contribution to the culture, as a monumental case study in humanity's most perplexing disease.

Six of the Galvin boys took ill at a time when so little was understood about schizophrenia—and so many different theories were colliding with one another—that the search for an explanation overshadowed everything about their lives. They lived through the eras of institutionalization and shock therapy, the debates between psychotherapy versus medication, the needle-in-a-haystack search for genetic markers for the disease, and the profound disagreements about the cause and origin of the illness itself. There was nothing generic about how they each experienced the illness: Donald, Jim, Brian, Joseph, Matthew, and Peter each suffered differently, requiring differing treatments and a panoply of shifting diagnoses, and prompting conflicting theories about the nature of schizophrenia. Some of those theories could be especially cruel to the parents, who often took the blame, as if they'd caused the disease by something they did or did not do. The entire family's struggle doubles as a thinly veiled history of the science of schizophrenia—a history that for decades took the form of a long argument over not just what caused the illness, but what it actually is.

The children who did not become mentally ill were, in many respects, as affected as their brothers. It is hard enough to individuate oneself in any family with twelve children; here was a family that was defined by dynamics like no other, where the state of being mentally ill became the norm of the household, the position from which everything else had to start. For Lindsay, her sister, Margaret, and their brothers John, Richard, Michael, and Mark, being a member of the Galvin family was about either going insane yourself or watching your family go insane—growing up in a climate of perpetual mental illness. Even if they happened not to descend into delusions or hallucinations or paranoia—if they didn't come to believe that the house was under attack, or that the CIA was searching for them, or that the devil was under their bed—they felt as if they were carrying an unstable element inside themselves. How much longer, they wondered, before it would overtake them, too?

As the youngest, Lindsay endured some of the worst of what happened—left in harm's way, directly hurt by people she thought loved her. When she was little, all she wanted was to be someone else. She could have left Colorado and started over, changed her name for real, assumed a new identity, and tried to scribble over the memory of all she went through. A different person would have gotten out as soon as she could and never come back.

And yet here Lindsay is at Point of the Pines, checking to see if the brother she once dreaded needs a heart examination, if he's signed all the forms that need signing, if the doctor has seen him enough. She does the same for her other sick brothers, too, the ones still living. With Donald, for the length of her visit today, Lindsay pays careful attention as he wanders the halls. She worries that he is not taking good enough care of himself. She wants the best for him.

In spite of everything, she loves him. How did that change?

THE ODDS OF a family like this one existing at all, much less one that remained intact long enough to be discovered, seem impossible to calculate. The precise genetic pattern of schizophrenia has defied detection; its existence announces itself, but fleetingly, like flickering shadows on the wall of a cave. For more than a century, researchers have understood that one of the biggest risk factors for schizophrenia is heritability. The paradox is that schizophrenia does not appear to be passed directly from parent to child. Psychiatrists, neurobiologists, and geneticists all believed that a code for the condition had to be there somewhere, but have never been able to locate it. Then came the Galvins, who, by virtue of the sheer number of cases, offered a greater degree of insight into the illness's genetic process than anyone imagined possible. Certainly no researcher had ever encountered six brothers in one family—full-blooded siblings, with the same two parents in common, the same shared genetic line.

Starting in the 1980s, the Galvin family became the subject of study by researchers on the hunt for a key to understanding schizophrenia. Their genetic material has been analyzed by the University of Colorado Health Sciences Center, the National Institute of Mental Health, and more than one major pharmaceutical company. As with all such test subjects, their participation was always confidential. But now, after nearly four decades of research, the Galvin family's contribution finally can be seen clearly. Samples of their genetic material have formed the cornerstone of research that has helped unlock our understanding of the disease. By analyzing this family's DNA and comparing it with genetic samples from the general population, researchers are on the cusp of making significant advances in treatment, prediction, and even prevention of schizophrenia.

Until recently, the Galvins were completely unaware of how they might be helping others—oblivious to how their situation had, among some researchers, created such a feeling of promise. But what science has learned from them is only one small portion of their story. That story begins with their parents, Mimi and Don, and a life together that took flight with limitless hope and confidence, only to curdle and collapse in tragedy, confusion, and despair.

But the story of the children—of Lindsay, her sister, and her ten brothers—was always about something different. If their childhood was a funhouse-mirror reflection of the American dream, their story is about what comes after that image is shattered.

That story is about children, now grown, investigating the mysteries of their own childhood—reconstituting the fragments of their parents' dream, and shaping it into something new.

It is about rediscovering the humanity in their own brothers, people who most of the world had decided were all but worthless.

It is about, even after the worst has happened in virtually every imaginable way, finding a new way to understand what it means to be a family.

THE GALVIN FAMILY

PARENTS

"DON"
DONALD WILLIAM GALVIN
born in Queens, New York, on
January 16, 1924
died on January 7, 2003

"MIMI"
**MARGARET KENYON BLAYNEY
GALVIN**
born in Houston, Texas, on November
14, 1924
died on July 17, 2017

CHILDREN

DONALD KENYON GALVIN
born in Queens, New York, on
July 21, 1945
married Jean (divorced)

JAMES GREGORY GALVIN
born in Brooklyn, New York, on
June 21, 1947
married Kathy (divorced), one child
died on March 2, 2001

JOHN CLARK GALVIN
born in Norfolk, Virginia, on
December 2, 1949
married Nancy, two children

BRIAN WILLIAM GALVIN
born in Colorado Springs, Colorado,
on August 26, 1951
died on September 7, 1973

"MICHAEL"
ROBERT MICHAEL GALVIN
born in Colorado Springs, Colorado,
on June 6, 1953
married Adele (divorced), two children
married Becky

RICHARD CLARK GALVIN
born in West Point, New York, on
November 15, 1954
married Kathy (divorced), one child
married Renée

JOSEPH BERNARD GALVIN
born in Novato, California, on August
22, 1956
died on December 7, 2009

MARK ANDREW GALVIN
born in Novato, California, on August
20, 1957
married Joanne (divorced)
married Lisa, three children

MATTHEW ALLEN GALVIN
born in Colorado Springs, Colorado,
on December 17, 1958

PETER EUGENE GALVIN
born in Denver, Colorado, on
November 15, 1960

**MARGARET ELIZABETH GALVIN
JOHNSON**
born in Colorado Springs, Colorado,
on February 25, 1962
married Chris (divorced)
married Wylie Johnson; daughters
Ellie and Sally

"LINDSAY"
MARY CHRISTINE GALVIN RAUCH
born in Colorado Springs, Colorado,
on October 5, 1965
married Rick Rauch; son Jack,
daughter Kate

Part One

CHAPTER 1

1951

Colorado Springs, Colorado

Every so often, in the middle of doing yet another thing she'd never imagined doing, Mimi Galvin would pause and take a breath and consider what, exactly, had brought her to that moment. Was it the careless, romantic tossing aside of her college education in favor of a wartime marriage? The pregnancies and the children, one after another, with no plan of stopping if Don had anything to say about it? The sudden move out west, to a place that was completely foreign to her? But of all the unusual moments, perhaps none compared to when Mimi—a refined daughter of Texas aristocracy, by way of New York City—clutched a live bird in one hand and a needle and thread in the other, preparing to sew the bird's eyelids shut.

She had heard the hawk before she saw it. It was nighttime, and Don and the boys were asleep in their new home when there was an unfamiliar noise. They had been warned about coyotes and mountain lions, but this sound was different, the pitch high, the quality otherworldly. The next morning, Mimi went outside, and on the ground, not far from the cottonwood trees, she noticed a small scattering of feathers. Don suggested they bring the feathers to a new acquaintance of his, Bob Stabler, a zoologist who taught at Colorado College, a short walk from where they were living in the center of Colorado Springs.

Doc Stabler's house was unlike any place they had seen in New York: a home that doubled as a repository for reptiles, mainly snakes, including one that was uncaged—a cottonmouth moccasin, coiled around the back of a wooden chair. Don and Mimi brought their three sons with them, ages six, four, and two. When one of the boys dashed in front of the snake, Mimi shrieked.

"What's the matter?" Stabler said with a smile. "Afraid it's going to bite your baby?"

The zoologist had no trouble identifying the feathers. He had been training hawks and falcons as a hobby for years. Don and Mimi knew nothing about falconry, and at first they feigned interest as Stabler went on about it: how, in medieval times, no one beneath the rank of an earl was even allowed to own a peregrine falcon; and how this part of Colorado was a prime nesting spot for the prairie falcon, a cousin of the peregrine and every bit as majestic, he said, a thing of beauty. And then, against their better judgment, both Mimi and Don found themselves fascinated, as if they were being let in on one of the great private worlds of a place they were only just beginning to understand. Their new friend made it sound like a cultish thing, an archaic pastime practiced today by a secretive few. He and his friends were taming the same sorts of wild birds once tamed by Genghis Khan, Attila the Hun, Mary Queen of Scots, and Henry VIII—and they were doing it very much the same way.

In truth, Don and Mimi may have come to Colorado Springs about fifty years too late. Back then, this part of the state had been an agreeable destination for, among others, Marshall Field, Oscar Wilde, and Henry Ward Beecher, all of whom came to take in some of the natural wonders of the American West. There was Pikes Peak, the fourteen-thousand-foot summit named for an explorer, Zebulon Pike, who never actually made it to the top. There was the Garden of the Gods, the looming natural arrangement of sandstone rock outcroppings that seem staged for maximum effect, like the heads on Easter Island. And there was Manitou Springs, where some of the wealthiest, most refined Americans came to partake of the latest pseudoscientific cures. But by the time Don and Mimi arrived, in the winter of 1951, the elite sheen of the place had long worn away, and Colorado Springs had gone back to being a drought-ridden, small-minded outpost of a town—such a tiny pinpoint on the map that when the Boy Scout international jamboree was held there, the jamboree was bigger than the town.

So for Don and Mimi to happen upon such a grand tradition right under their noses—the mark of nobility and royalty, right there, in the middle of nowhere—sent shock waves through them both, feeding into their shared love of culture and history and sophistication. They were goners. But joining that club took some time. Aside from Doc Stabler, no one was willing to talk about falconry with the Galvins. Falconry was so exclusive, it seemed, that conventional bird-watching groups of the time had yet to embrace the pursuit of these particular birds.

Mimi could never remember how, but Don got his hands on a copy of *Baz-nama-yi Nasiri,* a Persian falconry text that only in the past few decades had been translated into English. From that book, he and Mimi learned to build their first trap, a dome made of chicken wire, affixed to a circular frame the size of a hula hoop. Following the instructions, they staked a few dead pigeons inside the trap as bait, with wires of fishing line hanging from the chicken wire above. At the end of each line, they tied slipknots to catch any bird who fell for the ruse.

Their first customer, a red-tailed hawk, flew off, carrying the whole trap behind it; their English setter ran after it and tracked it down. This was the first wild bird that Mimi ever held in her hand. Like a dog chasing a fire truck, she had no idea what to do if she caught one.

Back to Doc Stabler she went, hawk in hand. "Well, you did pretty well," he said. "Now sew the eyelids together."

Stabler explained that a falcon's eyelids protect them as they dive at speeds upward of two hundred miles per hour. But in order to train a hawk or falcon the way Henry VIII's falconers did it, the bird's eyelids should be temporarily sewn shut. With no visual distractions, a falcon can be made dependent on the will of a falconer—the sound of his voice, the touch of his hands. The zoologist cautioned Mimi: Be careful the stitches aren't too tight or too loose, and that the needle never pricks the hawk's eyes. There seemed any number of ways to make a hash of the bird. What, again, brought Mimi to this moment?

She was frightened, yet not entirely unprepared. Mimi's mother had made dresses during the Depression—even ran her own business—and she had made sure her daughter knew a few things. As carefully as she could, Mimi went to work on the edge of each eyelid, one after the other. When she was done, she took the long tails of the threads from both eyes, tied them together, and stashed them in the feather on top of the bird's head, to keep the bird from scratching at them.

Stabler complimented Mimi on her work. "Now," he said, "you have to keep it on the fist for forty-eight hours."

Mimi balked. How could Don walk the halls of Ent Air Force Base, where he worked as a briefing officer, with a blinded hawk on his wrist? How could Mimi do the dishes, or look after three small boys?

They divided up the work. Mimi took days, and Don took nights, during his late shifts at the base, tethering the bird to a chair in the room where he spent most of his time. Only once did a senior officer walk in and cause the hawk to "bate"—a falconry term meaning to fly away in a panic. Classified documents went flying everywhere, too. Don had a reputation at the base after that.

But at the end of those forty-eight hours, Mimi and Don had successfully domesticated a hawk. They felt an enormous sense of accomplishment. This was about embracing the wild, natural world, but also about bringing it under one's control. Taming these birds could be brutal and punishing. But with consistency and devotion and discipline, it was unbelievably rewarding.

Not unlike, they often thought, the parenting of a child.

WHEN SHE WAS a little girl, Mimi Blayney would sit under her family's grand piano and listen to her grandmother playing Chopin and Mozart. On nights when her grandmother picked up the violin, Mimi would stare, transfixed, as her aunt danced like a Gypsy along to the music, the logs in the fireplace crackling loudly behind her. And when no one else was around, the pale, dark-haired girl, no older than five, would venture where she was not allowed to go. The Victrola was broken more often than it wasn't, and the records the family owned—thick, grooved platters, more like hubcaps than LPs—were

filled with music that Mimi was dying to hear. When the coast was clear, Mimi would put a platter on the machine, place down the needle, and spin it with her finger. She would get about two measures of opera that way, over and over again.

The excavation of levees had worked out well for Mimi's grandfather, Howard Bowman Kenyon, a civil engineer who, long before Mimi was born, founded a company that dredged the rivers of five states, building levees along the Mississippi. Mimi's mother, Wilhelmina—or Billy to everyone—went to a private school in Dallas, and when the teacher would ask, "And what does your daddy do?" she'd coyly reply, "He's a ditch digger." At its wealthiest, during the Roaring Twenties, the Kenyon family owned its own island at the mouth of the Guadalupe River near Corpus Christi, Texas, where Mimi's grandfather dug his own lake and stocked it with bass. Most of the year, the family lived in a grand old mansion on Caroline Boulevard in Houston. In the driveway were two Pierce-Arrows, a fleet that increased by one additional Pierce-Arrow each time one of Grandfather Kenyon's five children came of age.

Mimi grew up with plenty of stories about the Kenyons. In her later years, she would recite those stories to her friends and neighbors and everyone she met, like secrets too delicious to keep to herself. The family's first home in Texas was sold to Howard Hughes's parents. . . . Howard Hughes himself was a classmate of Mimi's mother at the Richardson School, the academic institution of choice for Houston's upper crust. . . . Obsessed with mining, Grandfather Kenyon once traveled to the mountains of Mexico in search of gold and was briefly held captive by Pancho Villa, until his command of the local geography impressed the Mexican revolutionary so much that the two men struck up a friendship. Out of insecurity or, maybe, just a restless intellect, Mimi would come back to these stories as a way of affirming her status, her pedigree. It felt good to remind herself that there was something special about where she came from.

It made sense, by Kenyon standards, that when Mimi's mother, Billy, found someone good enough to marry, the groom was not just a twenty-six-year-old cotton merchant; he was the son of a scholar who had traveled the world as a trusted advisor to the banker and philanthropist Otto Kahn. The families of Billy Kenyon and John Blayney were perfectly matched, and the young couple seemed destined for a life of high-minded adventure. They set up a home of their own and had two children: first Mimi, in 1924, and then her sister, Betty, two and a half years later. The family's first real crisis came in early 1929, when Mimi's father, who had failed to measure up to the reputation of his family in practically every important way, exposed Mimi's mother to gonorrhea.

Grandfather Kenyon went after his son-in-law with a rifle, securing a quick divorce for his daughter. Billy and the girls moved back to the family home in Houston. Billy was helpless, on the verge of despair. A divorced, scandalized mother of two little girls—Mimi was five, Betty three—was not going to build any sort of life in the circles that the Kenyon family traveled in. There didn't seem to be a solution to the problem—until, a few months later, Mimi's mother fell for an artist from New York.

Ben Skolnick was a painter who had been just passing through town, on his way to create a mural in California. Ben had good taste, and was raised in a family of creative people, but he stuck out a little in Houston, not just because of what he did for a living but because he was Jewish. Billy's parents made sure to meet with Ben out of town, where no one would see them. But when Ben proposed, Billy's mother encouraged her to accept. No matter what her family might have thought about Ben Skolnick personally or Jews generally, they understood that this was Billy's best hope.

In the summer of 1929, Grandfather Kenyon drove Mimi, her mother, and her little sister to a boat in Galveston, Texas, which took them east along the Gulf to New Orleans, where they boarded a cruise ship on the Cunard line to New York. On board, the future Mrs. Skolnick and her daughters received invitations to sit at the captain's table, where they were required to have perfect manners, finger bowls included. Mimi got seasick easily, and even when she was well, she failed to enjoy the trip. Not for the last time, Mimi wondered if anything about her life would ever be the same.

THE NEWLY CONSTITUTED family struggled right away. Ben couldn't find any murals to paint after the stock market crash. Billy, with her refined breeding and eye for fine fabrics, found a job at Macy's. In time, she started a dress business in Manhattan's Garment District that brought the family a little more stability. While she worked, Ben and his family tended to the girls in their tiny house in Bellerose, Queens—the city's edge, practically on the border with Long Island.

New York slowly grew on Mimi. Sack lunches in hand, she and her sister could take the bus and subway for a nickel from far-off Queens to the Metropolitan Museum of Art in Manhattan, then make their way through Central Park, past Cleopatra's Needle to the Museum of Natural History, and find their way back home before dark. All the New Deal WPA projects allowed Mimi to see theater in ballparks and high school auditoriums. School took her on her first trips to the aquarium and planetarium. Her first ballet, by Léonid Massine, was staged inside the Met. Mimi would never forget the sight of twelve little girls who came all the way from Russia to dance—all, it seemed at the time, just for her. If the first world Mimi had known was the world of the Victrola and the grand piano and the country club and Junior League of Houston, she took more fiercely to this new world. "I loved growing up in New York," she would say. "That's the best education in the world, it really is."

And in the years to come, whenever things seemed awry in her life, Mimi's stories about her charmed New York childhood and gilded Houston family would, all together, paper over the gloom. Grandfather Kenyon hit some hard times in the Depression and had to let the family's loyal servants go, but beneficently permitted them to stay on his property, rent-free. . . . Mimi and her mother once traveled to Texas on the same train as Charlie Chaplin, and she played with the Little Tramp's children (who were rascals in their own right). . . . In the 1930s, Mimi's mother, Billy, accompanied Grandfather Kenyon back to Mexico, where she went drinking with Frida Kahlo and shook the hand of her exiled Russian friend, Leon Trotsky. . . .

As far as Mimi was concerned, these stories were better than the one about how much Ben Skolnick liked to drink. Or how she never saw her real father, John Blayney, again, and how much that hurt. Or how deeply, achingly she had longed for a life that would be as safe and secure as it would be extraordinary.

MIMI MET THE man who would offer her that life in 1937, when they were both still practically children. Don Galvin was fourteen and tall and pale, with hair as dark as hers. She was a year younger, studious but also quick to laugh. They were at a swim competition, and she'd blown a start, diving in before the whistle blew, and he was sent in to bring her back. After the meet, Don asked her out. That was the first time such a thing had happened to Mimi. She said yes.

Don was a serious-minded boy, college-bound, a reader. All that appealed to Mimi. But he also was handsome in the most wholesome, all-American way: lantern-jawed with slicked-back hair, a matinee idol in the making. Don wasn't an extrovert, and yet when he opened his mouth, people seemed to listen. It was not so much what he said as how he sounded: Don had the voice of a crooner, practically singing everything he spoke, smooth and seductive. With that voice, one of his sons, John, later said, "he could hold you in the palm of his hand."

Mimi's mother was suspicious. There may have been some snobbery at play there. The Galvins were devout Catholics—a tribe as foreign to the Episcopal Kenyon family as a Jewish family would have been before Billy had met Ben. Don's father was an efficiency expert for a paper company, and his mother was a schoolteacher. Neither of these facts did much to impress Mimi's mother.

But there was snobbery on both sides. Don's mother noted how Mimi did all the talking in the relationship. Did that mean she would ride roughshod over her youngest son? And then came the refrain from both sides that dogged them for years: *You're both so young.*

Nothing seemed to convince them that they weren't meant for each other. It was true that their interests weren't completely in line: He loved the Dodgers; she loved the ballet. But when they were fifteen and sixteen, Mimi persuaded Don to take her to *Petrushka* featuring Alexandra Danilova, the ballerina who had left the Soviet Union with George Balanchine. When Don came home raving about the performance, his brothers teased him for days. In the summer, Billy took Mimi on a trip, ostensibly to see Grandfather Kenyon. The not-so-secret agenda was to get Mimi away from Don for a while. It didn't work: Mimi wrote Don letters all the way there and back. Once she came home, Don took her to see *The Wizard of Oz,* and the couple sang and skipped together all the way home. That fall, they went to dances together, and school basketball games and rallies and Friday night bonfires. That spring, they drove out together to warm-weather clambakes on Cedar Beach on Long Island's South Shore.

Slowly, everyone came around. As Don neared graduation, his parents invited Mimi and her family to dinner. The Galvins lived in a nicer house than Mimi's family's place, a Dutch Colonial with a vast living room covered by a deep, dark red Oriental rug. Billy took notice of this. From that point forward, Don became a welcome guest at Mimi's house on Friday nights to play Scrabble. On return visits to Don's house, Mimi would clown around with Don and his two brothers, George and Clarke, both of whom were as handsome as he was. Even Don's mother thawed a little when Mimi and Don visited the Cloisters, and Mimi wrote a school paper for Don on the tapestries there. Mimi was helping her son to better himself. That was all right with her.

Not everything about their romance was effortless. Every weekend, Don hosted dances as the grand master of the Sigma Kappa Delta fraternity. Mimi went broke making new dresses each week, determined not to let anyone else go with him. There was a price to pay, perhaps, for going steady with the boy the Jamaica High School paper once called "Senior Head School Romeo." *Nothing but an absolute refusal to discuss his affairs of the heart could be obtained from the very secretive and the shy Mr. Don Galvin.*

Something about him—not just his looks, but a relaxed, easygoing self-assurance—made him both irresistible and, in some strange way, unattainable. That air of mystery would work to Don's advantage for much of his life. From the very start, it was as if Mimi belonged to him, while he belonged to everyone.

MIMI LOVED DON for his ambition, even if in her heart of hearts she would have preferred that he stay close to home. After high school, he told Mimi he wanted to join the State Department and travel the world. In the fall of 1941, just a few months before Pearl Harbor, he enrolled at Georgetown University School of Foreign Service in Washington, D.C. A year later, Mimi enrolled in Hood College in Frederick, Maryland, to be closer to him. But it was only a matter of time before the war would catch up with them both.

In 1942, in the middle of his sophomore year at Georgetown, Don enlisted in the Marine Corps Reserve. The following year the Marines sent him to Villanova, Pennsylvania, for eight months of mechanical engineering training. Before completing the course, the trainees were offered a shortcut to the front lines: If they wanted to, they could transfer to the Navy right away, with a guaranteed admission to Officer Candidate School. Don took the deal. On March 15, 1944, he was off to Asbury Park, New Jersey, for Navy midshipman's basic training, and then to Coronado, California, where he awaited an assignment. In November, Don received his posting: He would serve as a landing craft operator on the USS *Granville,* a brand-new attack transport ship bound for the South Pacific. Don was going to war.

Not long before Christmas, just a few weeks before shipping out, Don called Mimi long-distance from Coronado. Would she visit? Mimi asked her mother for permission, and Billy said yes. As soon as Mimi arrived, she and Don drove to Tijuana and got married. After the briefest of honeymoons on the road, they returned to Coronado for a tearful farewell. It was during Mimi's long trip home, on a stop in Texas to see her Kenyon relatives, that she experienced morning sickness for the first time.

Their rapid-fire wedding suddenly made sense: During Don's last swing through New York, several weeks before she'd traveled west to be with him, Mimi and Don had conceived a child.

Don's parents, devout Catholics, were not satisfied with a Tijuana wedding. Before shipping out, their son secured a few days' leave and traveled across the country one more time. On December 30, 1944, Don and Mimi took their vows again, this time in the rectory of the Church of St. Gregory the Great in Bellerose, Queens. The next day, Don filled out a Navy form to change his next-of-kin from his parents to Mrs. Donald Galvin.

THE BRIDE SPENT months vomiting. Long, unresolvable bouts of morning sickness would be a hallmark of nearly all of Mimi's twelve pregnancies. Her young husband's ship approached Japan in May 1945, just in time for the climax of the American offensive in the Pacific. Don's role was to transport soldiers on small crafts from ship to shore. Listening to the radio for reports on the *Granville,* Mimi nearly fell apart when Tokyo Rose announced that Don's ship had been destroyed. That turned out to be wrong, but just barely.

Anchored near Okinawa, Don witnessed boats on either side of him being blown up by kamikazes. He spent hours dragging his dead comrades out of the water. Don would never discuss anything about what he saw or did, not with Mimi. But he survived. And on July 21, 1945, two weeks before the United States dropped the bombs that would bring an end to the war, Don received a telegram aboard the *Granville* from Western Union: IT'S A BOY.

CHAPTER 2

1903
Dresden, Germany

It makes a certain amount of sense that the most analyzed, interpreted, pored-over, and picked-apart personal account of the experience of being psychotically paranoid and wildly delusional would be
almost impossible to read.

Daniel Paul Schreber grew up in Germany in the middle of the
nineteenth century, the son of a renowned child-rearing expert of the
period who made a practice of turning his children into test subjects.
As a boy, he and his brother are believed to have been some of the first
people to experience Moritz Schreber's cold-water treatments, diets,
exercise regimens, and a device called the Schreber *Geradehalter,*
made of wood and straps, that was designed to persuade a child to sit
up straight. Schreber survived that childhood and grew up to be very
accomplished, first a lawyer and then a judge. He married and had
a family, and with the exception of a brief depression in his forties,
everything seemed just fine. Then, at the age of fifty-one, came his
collapse. Diagnosed in 1894 with a "paranoid form" of "hallucinatory
insanity," Schreber spent the next nine years near Dresden in Sonnen-
stein Asylum, Germany's first publicly funded hospital for the insane.

Those years in the asylum formed the setting—at least physically—of *Memoirs of My Nervous Illness,* the first major work about the mysterious condition then known as dementia praecox, and a few years later renamed schizophrenia. Published in 1903, this book became a reference point for practically every discussion about the illness for the next century. By the time the six boys of the Galvin family became ill, everything about how they would be viewed and treated by modern psychiatry was colored by the arguments about this case. In truth, Schreber himself hadn't expected his life story to attract much attention. He wrote the memoir mainly as a plea for his release, which explains why, at many points, he seems to be writing for an audience of one: Dr. Paul Emil Flechsig, the doctor who'd had him committed. The book starts with an open letter to Flechsig, in which Schreber apologizes for writing anything that the doctor might find too upsetting. There is just one small matter Schreber hopes to clear up: Is Flechsig the one who has been transmitting secret messages into his brain for the last nine years?

A cosmic mind-meld with his doctor—"even when separated in space, you exerted an influence on my nervous system," Schreber wrote—was the first of dozens of strange and miraculous experiences related by Schreber over more than two hundred pages. It also might have been the most coherent. In a manner decipherable, perhaps, to Schreber alone, he wrote passionately about the two suns that he saw in the sky and the time he noticed that one sun was following him around wherever he went. He devoted many pages to an impenetrable explanation of the subtle "nerve-language" that most humans didn't notice. The souls of hundreds of people, he wrote, used this nerve language to pass along crucial information to Schreber: reports of Venus being "flooded," the solar system becoming "disconnected," the constellation Cassiopeia about to be "drawn together into a single sun."

In this respect, Schreber had a lot in common with the oldest of the Galvin children, Donald, who, years later, would recite his Holy Order of Priests in front of seven-year-old Mary in their family's house on Hidden Valley Road. Like Donald, Schreber believed that what was happening to him wasn't just physical but spiritual. Neither he nor Donald nor any of the Galvins were observing their delusions at a remove, with a detached sense of curiosity. They were right there in it, thrilled and amazed and terrified and despairing, sometimes all at once.

Unable to free himself from his circumstances, Schreber did his best to bring everyone in there with him—to share the experience. Being in his universe could feel ecstatic one moment, then shockingly vulnerable the next. In his memoir, Schreber accused his doctor, Flechsig, of using the nerve language to commit something he called "soul murder" against him. (Souls, Schreber explained, were fragile things, "a fairly bulky ball or bundle" comparable to "wadding or cobweb.") Then came the rape. "Owing to my illness," Schreber wrote, "I entered into peculiar relations with God"—relations that, at first, seemed an awful lot like immaculate conception. "I had a female genital organ, although a poorly developed one, and in my body felt quickening like the first signs of life of a human embryo . . . in other words fertilization had occurred." Schreber's gender had transformed, he said, and he had become pregnant. While he might have felt touched by grace, Schreber instead felt violated. God was Dr. Flechsig's willing accomplice, "if not the instigator," of a plot to use his body "like that of a whore." Schreber's universe was, much of the time, an intense and frightening place, filled with horrors.

He had one grand ambition. "My aim," Schreber reflected, "is solely to further knowledge of truth in a vital field, that of religion." It didn't turn out that way. Instead, what Schreber wrote contributed far more to the emerging, provocative, and increasingly contentious discipline of psychiatry.

IN THE BEGINNING—BEFORE anyone turned the study of mental illness into a science and called it psychiatry—being insane was a sickness of the soul, a perversion worthy of prison or banishment or exorcism. Judaism and Christianity interpreted the soul as something distinct from the body—an essence of one's self that could be spoken to by the Lord, or possessed by the devil. In the Bible, the first portrait of madness was King Saul, who lost his mind when the spirit of the Lord departed him and was replaced by an evil spirit. In medieval France, Joan of Arc heard voices that were considered heretical, the work of Satan—an impression that was revised the other way, to be the voice of a prophet, after Joan's death. Even then, insanity's definition was a moving target.

For those looking even a little carefully, it was plain to see that madness sometimes ran in families. The most conspicuous examples involved royalty. In the fifteenth century, King Henry VI of England first became paranoid, then mute and withdrawn, and finally delusional. His illness formed the pretext for the power struggle that became the Wars of the Roses. He came by it honestly: His maternal grandfather, Charles VI of France, had the same condition, as did Charles's mother, Joanna of Bourbon, and Charles's uncle, grandfather, and great-grandfather. But it took until Schreber's lifetime for scientists and doctors to start talking about insanity as something biological. In 1896, the German psychiatrist Emil Kraepelin used the term *dementia praecox* to suggest that the condition started at an early age, unlike senility (praecox also being the Latin root of *precocious*). Kraepelin believed that dementia praecox was caused by a "toxin" or "connected with lesions of an as yet unknown nature" in the brain. Twelve years later, the Swiss psychiatrist Eugen Bleuler created the term *schizophrenia* to describe most of the same symptoms that Kraepelin had lumped into dementia praecox. He, too, suspected a physical component to the disease.

Bleuler chose this new word because its Latin root—*schizo*—implied a harsh, drastic splitting of mental functions. This turned out to be a tragically poor choice. Almost ever since, a vast swath of popular culture—from *Psycho* to *Sybil* to *The Three Faces of Eve*—has confused schizophrenia with the idea of split personality. That couldn't be further off the mark. Bleuler was trying to describe a split between a patient's exterior and interior lives—a divide between perception and reality. Schizophrenia is not about multiple personalities. It is about walling oneself off from consciousness, first slowly and then all at once, until you are no longer accessing anything that others accept as real.

Regardless of what psychiatrists began to believe about the biology of the disease, its precise nature remained hard for any of them to fathom. While it seemed enough, at first, to say that schizophrenia could be inherited, that failed to account for cases—including, it seemed, Schreber's—where it seemed to appear all by itself. This essential question about schizophrenia—does it run in families or emerge fully formed out of nowhere?—would consume theorists and therapists and biologists and, later, geneticists, for generations. How can we know what it is until we know where it comes from?

———

WHEN SIGMUND FREUD finally cracked open Schreber's memoir in 1911, eight years after it was published, what he read took his breath away. The Viennese analyst and theorist, already widely revered as a pioneering explorer of the internal workings of the mind, showed no interest in delusional psychotics like Schreber. He had seen such patients as a practicing neurologist, but he had never thought it was worth the trouble to put any of them on the analyst's couch. Having schizophrenia, he argued, meant that you were incurable—too narcissistic to engage in a meaningful interaction with an analyst, or "transference."

But this book by Schreber—sent to him by his protégé, the Swiss therapist Carl Jung, who had pleaded with Freud to read it for years—changed everything for Freud. Now, without leaving his armchair, Freud had intimate access to every single impulse of a delusional man's mind. What Freud saw there confirmed everything he already thought he knew about the workings of the unconscious. In a letter thanking Jung, Freud called the memoir "a kind of revelation." In another, he declared that Schreber himself "ought to have been made a professor of psychiatry and director of a mental hospital."

Freud's *Psycho-Analytic Notes on an Autobiographical Account of a Case of Paranoia (Dementia Paranoides)* was published in 1911 (the same year that Schreber himself died, tragically enough, after reentering the asylum in the wake of his mother's death). Thanks to Schreber's book, Freud now was convinced that psychotic delusions were little more than waking dreams—brought on by the same causes as everyday neuroses, and interpretable in the very same way. All the same symbols and metaphors that Freud had famously noticed in dreams, he wrote, were all right there in the memoir, plain as day. Schreber's gender switch and his immaculate conception, Freud argued, were about a fear of castration. Schreber's fixation with his psychiatrist, Dr. Flechsig, he concluded, had to do with the Oedipus complex. "Don't forget that Schreber's father was a doctor," Freud wrote, triumphantly connecting the dots. "The absurd miracles that are performed on him (Schreber) are a bitter satire on his father's medical art."

No one seemed more tied up in knots over what Freud wrote than Carl Jung. From his home in Burghölzli, Switzerland, Jung read an early copy and wrote his mentor at once, in March 1911, to say he found it "uproariously funny" and "brilliantly written." There was just one problem: Jung fundamentally disagreed with him. At the heart of Jung's objection was the question of the nature of delusional mental illness: Is schizophrenia something you're born with, a physical affliction of the brain? Or is it acquired in life, after one has become scarred somehow by the world? Is it nature or nurture? Freud stood apart from most other psychiatrists of his time by being sure that the disease was entirely "psychogenic," or the invention of the unconscious, which had most likely been molded or scarred by formative childhood experiences—quite often of a sexual nature. Jung, meanwhile, held a more conventional opinion: that schizophrenia was at least partially an organic, biological illness—a disease that was quite likely inherited from one's family.

The protégé and his mentor had been sparring about this on and off for years. But for Jung, this was the last straw. He told Freud that not everything was about sex—that sometimes people go insane for other reasons, maybe because it is just something they're born with. "In my view the concept of libido . . . needs to be supplemented by the genetic factor," Jung wrote.

In several letters, Jung made that same case again and again. Freud never took the bait; he did not respond, which Jung found infuriating. By 1912, Jung exploded. He got personal. "Your technique of treating your pupils like patients is a blunder," Jung wrote. "In that way you produce either slavish sons or impudent puppies. . . . Meanwhile you remain on top as the father, sitting pretty."

Later that same year, before an audience at Fordham University in New York City, Jung spoke out against Freud in public, specifically blasting his interpretation of the Schreber case. Schizophrenia, he declared, "cannot be explained solely by the loss of erotic interest." Jung knew that Freud would consider this to be heresy. "He went terribly wrong," Jung later reflected, "because he simply doesn't know the spirit of schizophrenia."

The great break between Freud and Jung took place largely over the issue of the nature of madness itself. Early psychoanalysis's greatest partnership was over. But the argument over the origins and nature of schizophrenia was only just beginning.

A CENTURY LATER, across the world, schizophrenia affects an estimated one in one hundred people—or more than three million people in America, and 82 million people worldwide. By one measure, those diagnosed take up a third of all the psychiatric hospital beds in the United States. By another, about 40 percent of adults with the condition go untreated entirely in any given year. One out of every twenty cases of schizophrenia ends in suicide.

Academia is filled with hundreds of papers about Schreber now, each venturing far from Freud and Jung with their own takes on the patient and the illness that tormented him. Jacques Lacan, the French psychoanalyst and godfather of post-structuralism, said that Schreber's problems sprang from his frustration with somehow not being able to be the phallus that his own mother lacked. By the 1970s, Michel Foucault, the French social theorist and countercultural icon, held up Schreber as a sort of martyr, a victim of social forces working to crush the individual spirit. Even today, Schreber's memoir continues to be the perfect blank canvas, and Schreber himself is the ideal psychiatric patient: one who cannot talk back. Meanwhile, the central argument about schizophrenia raised by the Schreber case—nature or nurture?—has been baked into our perception of the disease.

This is the argument the Galvins were born into. By the time the Galvin boys came of age, the field was splitting open and dividing and subdividing almost like a cell. Some said the problem was biochemical, others neurological, others genetic, still others environmental or viral or bacterial. "Schizophrenia is a disease of theories," the Toronto-based psychiatric historian Edward Shorter has said—and the twentieth century produced easily hundreds of them. All the while, the truth about what schizophrenia was—what caused it, and what might alleviate it—has remained locked away, inside the people with the condition.

Researchers seeking a biological key to schizophrenia have never stopped searching for a subject or an experiment that might settle the nature-nurture question once and for all. But what if there was a whole family of Schrebers—a perfectly self-contained group with a shared genetic legacy? A sample set, with enough incidence of the disease that it seemed clear that something specific and identifiable must be happening inside some or even all of them?

A family like the dozen children of Don and Mimi Galvin?

CHAPTER 3

In the early years of their marriage, Mimi liked to joke that her husband would come home just long enough to get her pregnant.

Their first boy, Donald Kenyon Galvin, was baptized in September 1945, within a few days of Japan's surrender. His mother had endured his arrival in the world without incident; Donald's birth would be the only time Mimi would accept anesthesia for the birth of any of her twelve children. The baby and his mother lived together in a little apartment in Forest Hills, Queens, a peaceful section of New York City near the famous tennis club. Between strolls with the baby, Mimi taught herself to cook. For six months, she was alone with little Donald, listening to news reports from the South Pacific, wondering when her son's father would make it home.

Don returned just after Christmas, moved in with his family, and spent a few months on temporary duty as a security officer at a shipbuilding facility in Kearny, New Jersey. Then he was gone again, off to Washington for three months to finish his bachelor's degree at Georgetown. And then, in the summer of 1947, to the Navy's General Line School in Newport, Rhode Island—just weeks after Mimi gave birth to their second son, Jim. Don took Mimi and the kids with him that time, and again, a year later, to Norfolk, Virginia, where he served first on the USS *Adams,* and then on the USS *Juneau,* shuttling between New York and Panama, Trinidad, Puerto Rico, and the rest of the Caribbean—all while Mimi stayed home alone with the boys for weeks at a time.

Mimi had been nursing an entirely different dream of their life after the war. She had envisioned her husband going to law school, like her two uncles and her paternal grandfather, Thomas Lindsey Blayney, whom she adored despite her father's exile from the family. Mimi wanted to be in New York, where their families were, where their children would grow up with their cousins and aunts and uncles—a childhood like the one that had been ripped away from her when she was forced to move from Texas as a child.

Don had entertained that idea, or he seemed to. But he had dreams, too. He explained in his usual charming way that the Navy was a means to an end for him—that he thought he could get the Navy to sponsor his studies in the law or, better yet, his real passion, political science. This turned out to be a frustrating miscalculation. Despite glowing reviews and hearty recommendations from his commanding officers, he was turned down each time he applied for graduate-level course work. It always seemed that someone with connections, a congressman's son or a senator's nephew, got the appointment instead.

Alone in Norfolk, Mimi had to pinch pennies while Don was at sea. The small checks from the Navy, about thirty-five dollars a week, would get lost in the mail, and she would have to rely on her neighbors for groceries and meals. It was a different story when Don was in port. With his Georgetown education, his command of languages, and his interest in international relations, the handsome young lieutenant was making a good impression. Aboard the *Juneau,* Don wasn't just the ship secretary; he was the resident chess master, taking on all comers. Between missions, Don was the captain's regular tennis partner, and he and Mimi socialized with the brass at the Armed Forces Staff College in Norfolk, where Don became known for fixing Iron Curtains, a potent lowball made with vodka and Jägermeister. Don's smooth, professorial air impressed a number of admirals and generals—as well as at least one of their wives, who happened to come along as a passenger on one of the *Juneau*'s trips to Panama.

There aren't many places to find privacy on a warship, but there are enough. Back on the mainland, however, secrets are not so easy to keep. The officer's wife might not have known that one of her friends was acquainted with the wife of Don Galvin. When Mimi heard about that voyage of the *Juneau,* any last bit of allure of being the bride of a distinguished Navy lieutenant quickly faded. No one may have been more in thrall to Don than Mimi. But now, with two little boys to care for, she was all too aware that she needed him more than he needed her.

DON APPLIED TO a law program in exchange for committing to stay in the Navy another six years. He was turned down. He requested transfers to Panama, Cuba, or the Atlantic Division—all places where the Navy offered law classes. He was turned down again.

There was another violently ill pregnancy, followed by another son: their third, John, born in Norfolk at the end of 1949. Don was away for this one, in the middle of a deployment in Glenview, Illinois, for four months of officer's training. Mimi and the boys stayed in Norfolk as Don worked to be transferred somewhere, anywhere, else. Then Don received word that the *Juneau* was moving its home shipyard to Puget Sound—across the country on the West Coast, one step closer to Korea, where war was brewing.

Mimi couldn't contain herself any longer. It was time for Don to leave the Navy. On January 23, 1950, Don gave notice in a letter that laid the blame squarely on his home situation. "Deprivation of a wholesome family life is reason enough for my resignation," Don wrote. "To remain in the Navy would deprive my wife and my three sons of a normal family life and a home." Don also appeared to be stinging from his rejections—all the moments when the Navy had failed to recognize his potential. He'd had enough of being passed over for law school. "Motivation," he wrote, "can only come when we want to do something, or someone instills in us a desire to do it. I have experienced no motivation in the Navy."

Mimi was relieved. Finally, her long exile in strange, faraway towns would come to an end. They planned to move back to New York, where Don would enroll at Fordham Law School in the Bronx, and they would get started on the life she'd wanted all along. They

shopped for a house in Levittown, Long Island's new enclave of af-
fordable mass-produced houses within driving distance of the city,
and they set their sights on a place large enough for little Donald and
Jim and John, plus whoever else might come along.

What Mimi did not know was that Don also had been talking
with his brother Clarke, who had recently become an officer in the
United States Air Force. Unlike the Navy, everything about the Air
Force was still fresh and unformed. The pilots didn't even have the
blue uniforms yet, just the khaki "pinks and greens" left over from
its wartime incarnation as the Army Air Corps. And they seemed to
need people badly—so much so that Don learned that if he joined,
they'd make him an officer instantly.

On November 27, 1950, ten months after he'd left the Navy,
Don joined the Air Force as a first lieutenant. Mimi could not believe
how blithely Don seemed to be reneging on every understanding she
thought they had about how they wanted to live their lives. America
was sending troops to Korea, and he wanted back in? Why was he
always one half step out of sync with her—so remote, so absent?

Don was as persuasive with Mimi as ever. Clarke had taken him
out one day to see Mitchel Field, the air base on Long Island that was
serving as the military branch's national headquarters. Did it really
matter to Mimi, he asked, whether he was commuting to the Bronx
to study law or Long Island to train? Either way, they could still live
in Levittown. Besides, Don still had dreams. America was leading the
world now, building the future. The air fleet that had just defeated
fascism would be flying in and out of his and Mimi's backyard. Did
he want to push paper in some skyscraper and catch the 5:07 home
every night? Or did he want to be a part of that—an expert in international affairs someday, with the ear of presidents?

Mimi and Don put together enough money for a deposit on a
house. They had almost closed on the place when the Air Force announced, quite suddenly, that its new headquarters would be in the
middle of the state of Colorado. This time, Don was as shocked as
Mimi was. The relocation had been planned behind the scenes in
Washington. No one they knew had known anything about it.

After a brief panic, they got their deposit back. Don reported to
Ent Air Force Base in Colorado Springs on January 24, 1951. Mimi
and the children joined him by Valentine's Day.

———

THERE WAS ROCK everywhere Mimi turned—miles of it, all different shades of red, tremendous open prairies pressed flat by glaciers and punctuated by violent outcroppings that towered over the flatlands like a stage set. There were the spas of Manitou Springs, spouting mineral water said to possess amazing healing powers. And the mountains where the previous century's gold rush had first put this part of Colorado on the map. Beauty surrounded Mimi, even if she was in no mood to see it.

The town was not looking its best when they got there. Mimi and the boys had arrived in the middle of a drought. Water was being rationed. Even Mimi's mother's house in New York City had green grass and flowers; now everywhere Mimi looked, she saw brown. There was no ballet and no art or culture here—nothing close to the life that Mimi had dreamed of as a girl. The house Don found for them was located on what passed for a bustling boulevard in Colorado Springs, a silent street called Cache La Poudre. This was about as different from Levittown as a person could imagine: an old converted feed barn with a stairway with floorboards that were hopelessly bowed and crooked.

Mimi cried for several nights and seethed for longer than that. The house was a dump, she said, the town a backwater. Where exactly had he dragged her now?

But Don was her husband. And she was a mother of three, with plans for more—Don was a Catholic, after all—and plenty to do no matter where she was. Mimi decided to try to make the most of it. The birds helped—the Oregon juncos and the gray-crowned rosy finches and the mountain chickadees. There was a big cottonwood tree in the yard, and when she stared a little closer at the brown dirt, she saw wildflowers. She decided that she would plant a garden there.

Mimi's new neighbors on Cache La Poudre came to know her as a conspicuous reader of very thick books, a woman who could recite the names of every king and queen not just from Great Britain but from every country in Europe, from the Dark Ages until the present day. They soon learned all about Grandfather Kenyon and Pancho Villa and Howard Hughes and her years in New York. And on her husband's modest income, Mimi searched for other ways to seem special. From her mother, Mimi knew everything there was to know about the best fabrics, so she would scope out a bit of cashmere that had found its way into the Goodwill and then crow about her catch. She found a local choir to sing in and volunteered as an organizer with an amateur opera group. They wouldn't stage anything by her favorite, Mozart, at first—even *that* was too challenging for them, she'd scoff privately—but Mimi helped with the casting of performers for *Il Trovatore* and *Madama Butterfly,* all the old standards.

In time, she came to love the beauty around her. The plants and geology, all so foreign to Mimi, now made it seem as if everything she had once gazed at through glass in the Museum of Natural History on Central Park West was coming alive before her eyes. And together with Don, she discovered falconry. The cultivation of such feral birds managed to blend the intense intellectual might on which they'd built their relationship with something wild and undiscovered, like their new home.

Training a falcon, they both learned, was more than just trapping; it was also about the relentless imposing of one's will—maintaining control until the bird develops a sort of Stockholm syndrome, agreeing to stick around and even preferring captivity to being out in the world. After two weeks carrying the blinded bird on a gloved fist, or gauntlet, they would tie a creance—a one-hundred-foot string the same weight as fishing line—to the bird to maintain control during training. With some meat in a leather pouch to lure him home, they encouraged the bird to fly farther and farther away—until finally, they swung the lure out of reach to teach the bird to make diving passes. Diving, as they were deeply thrilled to witness, at upward of two hundred miles per hour.

As tricky as it was, the method for domesticating a wild hawk or falcon was well articulated—and if followed correctly, she and Don learned, you ended up with a well-behaved, obedient, civilized bird. Mimi also applied this persistent, unyielding approach at home, where sometimes there were more allowances made for the birds than for the children. The garage shelves were filled with leather hoods for the birds, and eventually the garage itself became a mews. (When one neighbor called the board of health on them, Don, who kept a clean mews, fended them off easily.) Mimi had taken a cheap watercolor set and started painting renderings of falcons. And together they introduced their new obsession to their boys. When Donald, their oldest, was grade-school age, he took part in the trapping of his first bird, a female sparrow hawk. They found her in a hole in a tree while bird-watching in Austin Bluffs, a 6,600-foot summit that once was the home of a tuberculosis sanitarium and would one day be the site of a University of Colorado campus. Mimi named the hawk Killy-Killy, after the *killeee* cry she made. Donald trained her himself. Once she caught a grasshopper and flew up to the top of a door, and started nibbling at the grasshopper like an ice cream cone. Donald stood below the door, patiently calling out, "Come Killy-Killy! Come Killy-Killy!" Back in the house, he'd let Killy-Killy fly loose, and they learned to step out of the way whenever she tilted her tail a certain way to poop.

The oldest two boys, Donald and Jim, started school. While the third boy, John, was still a toddler, the fourth and fifth, Brian and Michael, were born in 1951 and 1953. As infants, all the boys were breast-fed, a less-than-popular choice among most of the mothers Mimi knew. From the start, she felt good showing everyone that she could do everything on her own—no nannies, no baby-sitters. Who needed anyone else, Mimi thought, when she obviously was the best person to teach the boys, as they grew older, about opera and art and the observation of exotic birds, the examination of strange insects, and the identification of wild mushrooms? How many other children in Colorado Springs knew that the red polka-dotted ones were *Amanita muscaria*?

One after another, each boy got the mumps, the measles, and the chicken pox. With each new baby, the competition for Mimi's attention increased, as did the demands on her time. Even with five boys, neither Don nor Mimi made any mention of stopping. The refrain from both sides of the family was ceaseless: *Why so many children?* After all, Mimi's attraction to the finer things in life—culture, art, social status—hardly seemed compatible with having so many mouths to feed. But if Mimi couldn't have the former, she was more than happy to try her hand at the latter. There was a different sort of distinction in having so many children, and being known as a mother who could easily accomplish such a thing.

At the same time, no amount of social ambition could explain everything about Mimi's desire for a large family. There was quite likely another, deeper explanation as well—that the children filled a need in Mimi that perhaps even she had not anticipated. From an early age,

Mimi had a way of glossing over the more painful disappointments in her life: the loss of her father; the forced exile from Houston; the husband who remained so distant from her. Even if she didn't admit it, these losses hurt, and took their toll. Having so many children, however, offered Mimi a brand-new narrative—or at least distracted her, changed the subject, shored up the losses, helped her dwell less on what was missing. For a woman who so often felt abandoned, here was a way to create all the company she would ever need.

Don's mother, Mary Galvin, holding forth from her home in Queens, would say, somewhat cruelly, that the pregnancies were all Mimi's doing—that Mimi ran Don's life now, and that Mimi wanted the upper hand in all things, and that she was determined to out-Catholic the real Catholics in the family, and Mimi's perpetual state of pregnancy was the clearest and most powerful way to win that competition.

For Mimi, the response to that was simple, stopping all conversation. The children, she said, made Don happy.

HE WAS ALWAYS more of a scholar than a soldier. Mimi found that part of Don both lovable and frustrating. At the same time that he insisted on having a house filled with children, he also treasured a life of the mind, of solitude and order. And yet no matter how tranquil and orderly she made their home, he always found a reason to stay away.

As an intelligence officer at Ent Air Force Base, Don embraced the circumspect nature of Cold War military work. "Don't give anyone any more information than you have to," he used to say, and his coy way of saying it made the air of secrecy seem almost conspiratorial, something they all shared. They didn't share it, though: The most Don would confide in Mimi was that the generals he was briefing didn't seem terribly bright. Despite how well he seemed to be doing there, his ambition as an Air Force man had limits. Even when President Dwight Eisenhower set up his summer White House in Denver in 1953, and Don found himself drafting the intelligence briefings that Eisenhower himself was reading, military work interested Don only insofar as it made him even more determined to get a PhD in political science one day.

Where falconry once was something Don and Mimi did together, that started to change. He spent more time away from home, luring birds with other local falconers, while Mimi's work caring for the children was never-ending. This new disconnection maybe wasn't so new—more likely, it revealed something about them that had been the case from the very beginning. From the first day they'd met, Don had always seemed to be living his life a few inches off the ground, while Mimi had waited patiently, her feet planted firmly on earth. Don identified with his birds—soaring where he pleased, returning only when it suited him. And Mimi, quite against her will, found herself cast in the role of falconer—domesticating Don, luring him home, laboring under the impression that she had completely tamed him.

Mimi found her own ways of occupying herself, some designed to bring her closer to a husband who was growing further and further away from her. Fulfilling a promise she had made to Don's family, she went through several years of instruction to convert to Catholicism. Being the same religion as her husband made their family a real family, and so she did this happily—another mountain to climb, another subject to master. She formed a lasting friendship with her tutor, Father Robert Freudenstein, a local priest who introduced her to concepts like transfiguration and the virgin birth, all over cocktails. This was Mimi's kind of priest: Freudy, as he was called, came from some money and wasn't afraid to show it off, driving his convertible so fast that the birds outside their house would scatter when he pulled up. Freudy performed sleight of hand tricks for the boys and told them stories. With Mimi and Don, he talked about books and art and music, helping them feel less alien in their new home. When the Royal Ballet came to Denver, he took Mimi and Don together. Soon Freudy was dropping in at all hours, almost like another member of the family, whenever he needed to get away from his bosses at St. Mary's parish. "Oh, Monsignor Kipp is mad at me," he'd say. "Can I have breakfast with you?" Mimi always said yes.

Mimi's mother questioned the wisdom, and even the propriety, of this friendship. Billy would drive out west by herself in her Studebaker, and she'd stay until she started making comments about the way Mimi ran her household. Freudy was often Topic A. Marrying a Catholic was one thing, Billy would say, but must there always be a priest hanging around the house? But for Mimi, Freudy's visits were the most delightful surprise of her conversion to Catholicism. Not only could she become closer to Don and feel equipped to lead the spiritual training of her family, she had found something familiar, even fun, in what sometimes could be a lonely new existence.

Growing fed up, Billy would turn around and leave. But her mother's judgment only bothered Mimi a little. She had more children than Billy now. She outranked her.

THE MORE CHILDREN she had, the more Mimi grew into her new self—a different woman from the one who had been so disappointed for so many years. There would be other moves in their future: an Air Force transfer to a base in Quebec in 1954 and 1955, followed by three years at Hamilton Air Force Base in northern California. They returned to Colorado Springs in 1958 with eight boys. Richard was born in 1954, Joe in 1956, and Mark in 1957.

Don, when he was at home, was the good cop, a subdued presence, except for each morning at sunrise: *Reveille, reveille! Up all hands, heave out and trice up! Sweep down all decks and ladders fore and aft, report to the mess hall at 0600 for chow!* The rest of the time, it was Mimi who provided all supervision—not always nurturingly, but coolly, sharply, haughtily. She was a happy warrior, doing battle with mediocrity morning, noon, and night.

All the boys wore sport coats and ties and Bass Weejuns to Sunday mass.

Long hair was unacceptable.

The military and the Church supplied two sets of rules to follow: America's and God's.

Mimi was the master of every aspect of her children's lives, an endeavor in which she left absolutely nothing to chance. The children were raised on a bevy of axioms: "Pretty is as pretty does"; "Tattle Tale Tit, your tongue shall be split, and all the dogs in the town shall have a little bit." In the morning, everyone had their assignment: set the table, prepare lunch, make the toast, vacuum, dust, and mop the kitchen floor, clear the table, wash and dry the dishes. The assignments switched from week to week. The boys were enrolled in speed-reading classes. In good weather, they'd go out bird-watching or looking for mushrooms. Their living room had no issues of *Reader's Digest* or *Ladies' Home Journal*—only *Smithsonian* and *National Geographic*. Even the neighborhood children, when they came over to the Galvins' to color or draw or paint, learned to expect to hear not praise for their artwork but a detailed explanation of everything they were doing wrong. "She wanted everybody to be perfect," one old friend of the family remembered.

Mimi couldn't have known at the time how terribly this temperament would end up working against her. By the 1950s, the psychiatric profession had set its sights on mothers like her. The most influential thinkers in American psychiatry all were using a new term for such women. They called them "schizophrenogenic."

CHAPTER 4

1948
Rockville, Maryland

The Chestnut Lodge psychiatric hospital opened in 1910 in a modest, four-story brick building, once a hotel, on a tree-filled country estate in the outer reaches of Washington, D.C. For its first twenty-five years, the patients, many diagnosed with schizophrenia, were treated mostly with rest and occupational therapy; the hospital's founder lived downstairs while the patients lived upstairs. If few people in psychiatry thought much of the place, that all changed in 1935, when the hospital welcomed a new therapist named Frieda Fromm-Reichmann.

She had just arrived in America, a Jewish refugee from war-torn Germany. Already in her mid-forties, Fromm-Reichmann had established herself before the war as an experienced and confident psychotherapist—small, but forceful, intense, and direct—and the ideas she brought with her were undeniably fresh. Unlike some of the old-timers at Chestnut Lodge, Fromm-Reichmann was a member of a new wave of analysts who were inspired by Freud and willing to dare greatly with their patients. And before long, stories circulated about the miracles she was working.

There was the young man who assaulted Fromm-Reichmann when she first tried to talk to him. She held vigil outside his door daily for three months until he finally invited her in.

There was the man who kept silent for weeks during his sessions with Fromm-Reichmann, until one day he slipped a newspaper on the spot where she was about to sit. His first words to the doctor were something about not wanting her to dirty her dress.

And there was the woman who threw stones at Fromm-Reichmann, shouting, "God damn your soul to hell!" After a few months, the new therapist called her bluff. Clearly no one was benefiting from this, she said. "Why not stop it?" So the woman did.

Too good to be true? Perhaps. But to Fromm-Reichmann, schizophrenia was curable, and anyone who said differently might not care enough about the people they were treating. No member of the Galvin family ever met her. But no other person may have done more to change the way that schizophrenia and all mental illness was perceived in America during their lifetimes—for better and, later on, for worse.

FROMM-REICHMANN HAD ARRIVED in America at a moment when mainstream psychiatry's approach to schizophrenia was as ineffective as it was inhumane. Insane asylums were filled with test subjects who were forced to take cocaine, manganese, and castor oil; injected with animal blood and oil of turpentine; and gassed with carbon dioxide or concentrated oxygen (the so-called "gas cure"). The gold standard of treatment, in the 1930s, had been insulin shock therapy, in which the patient was injected with insulin to induce a short coma; the theory was that regular treatments, a coma a day, might slowly chip away at the effects of psychosis. Then came the lobotomy, the severing of the nerves of a patient's frontal lobes—which, as the British psychiatrist W. F. McAuley delicately put it, "deprives the patient of certain qualities with which, and perhaps because of which, he has failed to adapt."

Their counterparts searching for the biological cause of schizophrenia weren't treating their patients any better. In Germany, Emil Kraepelin, the dementia praecox pioneer, had opened an institute to research a hereditary link to the disease and had turned up little to nothing. A researcher at his institute, Ernst Rüdin, became a major

figure in the eugenics movement, among the first to argue for sterilizing the mentally ill. A student of Rüdin's named Franz Josef Kallmann went even further: Preaching eugenics in the United States after
the war, Kallmann called for sterilizing even "nonaffected carriers" of
a gene for schizophrenia, once such a gene was found. The leadership
of biological psychiatry seemed settled on the idea that mentally ill
people weren't people at all.*

In the face of such troubling social forces, it's hardly surprising
that Freud-inspired analysts like Fromm-Reichmann rejected the idea
of a biological basis of schizophrenia completely. Why should psychiatry sign on to a scientific discipline that treated humans like horses
to be selected for breeding? Instead, Fromm-Reichmann believed that
patients, deep down, wanted a cure—that they were waiting to be
helped, like a wounded bird or a frail child in need of understanding.
"Every schizophrenic has some dim notion of the unreality and loneliness of his substitute delusionary world," she wrote. And the therapist's mission—a high-minded undertaking that a new vanguard of
American psychoanalysts soon embraced—was to break through the
barriers the patient had erected and save them from themselves.

In 1948, Chestnut Lodge admitted a teenage girl named Joanne
Greenberg, who would go on to bring Fromm-Reichmann a measure of immortality. Greenberg's 1964 best-seller, *I Never Promised
You a Rose Garden*—a fictionalized memoir, she later called it—was
the story of a teenage girl named Deborah Blau who is trapped in the
delusional kingdom of Yr. Deborah believes herself to be possessed by
an outside force, much the way Daniel Paul Schreber felt that he had
been, a half century earlier. ("There were other powers contending for
her allegiance," Greenberg writes.) Deborah seems walled off from
the world forever until her therapist, Dr. Fried—a thinly disguised
Fromm-Reichmann, with a surname unmistakably echoing *Freud*—
breaks through and rescues her. Dr. Fried understands young Deborah's demons—their source and their reason for being. "The sick are
all so afraid of their own uncontrollable power!" Dr. Fried muses in

---

* The idea of sterilizing the insane and "feeble-minded" had caught on in America many years earlier. Eugenics was a hallmark of the turn-of-the-century Progressive Era in America, influencing Kallmann and Rüdin and, among others,
the Nazis.

the novel. "Somehow they cannot believe that they are people, holding only a human-sized anger!"

What Dr. Fried does for Deborah in this book influenced a generation of psychotherapists. Like Annie Sullivan in *The Miracle Worker*, Dr. Fried was a model of insight, compassion, and drive—patiently, ardently connecting with her patient, cracking her particular code. One of the keys, the doctor concludes, is recognizing that the girl's own parents had unwittingly fanned the flames of mental illness in their daughter. "Many parents said—even thought—that they wanted help for their children, even to show, subtly or directly, that their children were part of a secret scheme for their own ruin," the doctor reflects in the pages of Greenberg's novel. "A child's independence is too big a risk for the shaky balance of some parents."

The mystery of schizophrenia is, apparently, solved: The eugenicists are wrong. People aren't born with schizophrenia at all. Their mothers and fathers are to blame.

AS EARLY AS 1940, Fromm-Reichmann had sounded the alarm over "the dangerous influence of the undesirable domineering mother on the development of her children," calling such mothers "the main family problem." It was eight years later, the same year that Joanne Greenberg became her patient, that Fromm-Reichmann came up with a term that would stick to women like Mimi Galvin for decades: the schizophrenogenic mother. It was "mainly" this sort of mother, she wrote, who was responsible for the "severe early warp and rejection" that rendered a schizophrenia patient "painfully distrustful and resentful of other people."

She was far from the first psychoanalyst to blame the mother for something. Freud's approach, after all, was to explain practically every mysterious impulse as the end result of childhood experiences coloring the unconscious mind. But now, in the postwar years, the dawn of a new era of American prosperity, many therapists had something new to worry about: mothers who refused to behave like the mothers of a previous generation. "A schizophrenic," a Philadelphia psychiatrist named John Rosen wrote, within a year of Fromm-Reichmann's invention of the term schizophrenogenic mother, "is always one who is reared by a woman who suffers from a perversion of the maternal instinct."

In her own writings, Fromm-Reichmann remarked with unease at how "American women are very often the leaders, and men wait on them as wives wait on their husbands in European families," and how "the wife and mother is often the bearer of authority in the family group." She particularly disliked how fathers, like Don Galvin, become the confidants and pals of their kids, while mothers, like Mimi Galvin, become the disciplinarians. But once Fromm-Reichmann gave such mothers a name, the concept caught fire. John Clausen and Melvin Kohn from the National Institute of Mental Health described the schizophrenogenic mother as "cold," "perfectionistic," "anxious," "overcontrolling," and "restrictive." The psychologist Suzanne Reichard and the Stanford psychiatrist Carl Tillman described the schizophrenogenic mother as the "prototype of the middle class Anglo-Saxon American Woman: prim, proper, but totally lacking in genuine affection."

These descriptions seemed to lack a certain coherence. What, precisely, were these mothers doing to these children? Were they domineering or weak? Suffocating or withholding? Sadistic or apathetic? In 1956, Gregory Bateson, an anthropologist—and the husband of Margaret Mead—collected the various alleged sins of the schizophrenogenic mother into a theory he called the "double-bind." The double-bind, he explained, was a trap that certain mothers set for their children. A mother says, "Pull up your socks," but something about the way she says it projects the contradictory message, "Don't be so obedient." Now, even if the child obeys, the mother disapproves. The child feels helpless, frightened, frustrated, anxious—ensnared, with no way out. According to the double-bind theory, if children get caught in that trap often enough, they develop psychosis as a way of coping with it. Tormented by their mothers, they retreat into a world of their own.

Bateson invented this theory without so much as ten minutes of clinical psychiatric experience. But that made no difference. The double-bind, along with the schizophrenogenic mother, helped to turn mother-blaming into the industry standard for psychiatry—and not just for schizophrenia. In the 1950s and 1960s, it became hard to find any emotional or mental disorder that was not, in one way or another, attributed by therapists to the actions of the patient's mother. Autism was blamed on "refrigerator mothers" who failed to show enough affection to their infants. Obsessive-compulsive disorder was blamed on problems in the second to third year of life, clashing with the mother around toilet training. The public conception of madness became hopelessly intertwined with the idea of the mother-as-monster. When, in 1960, Alfred Hitchcock's *Psycho* placed the blame for the most famous delusional homicidal maniac of cinema, Norman Bates, squarely on the shoulders of his dead mother, it made all the sense in the world.

THIS IS WHAT the Galvins would be up against when their boys started getting sick: an emboldened therapeutic profession seizing the moral high ground, doing battle with the devils of eugenics and surgery and chemical experimentation, and more than ready to search for a different way to explain the disease—a cause much closer to home. In 1965, Theodore Lidz, a prominent Yale psychiatrist best known for attributing schizophrenia to a patient's family dynamics, said that schizophrenogenic mothers "became dangerous figures to males," and had "castrating" relationships with their husbands. As a general rule, Lidz recommended that schizophrenia patients be removed from their families entirely.

Parents of Don and Mimi Galvin's era didn't have to know about the double-bind theory or the schizophrenogenic mother to understand that anything wrong with their children would raise questions about them. What happened to those children when they were in their care? Who let them become this way? What sort of parents were they? The lesson of the times was clear. If something seemed off about your child, the last thing you should do is tell a doctor about it.

CHAPTER 5

When, after four years of out-of-town postings, the Galvins returned to Colorado Springs in 1958, the dusty town they'd left behind was fading into history. The United States Air Force Academy had opened while they were gone, and thousands of newcomers—cadets and their instructors and all the personnel needed to support a vast new military institution—were swiftly changing the character of the place. Where once there had been a dirt road with a couple of ruts, crossed by barbed wire gates that you had to open and close yourself, now there was Academy Boulevard, paved and leading to a gate that was guarded like it was the checkpoint between East and West Berlin. Inside, the Academy had its own post office, commissary, and telephone exchange. And the glistening new structures of the Academy itself were modernist masterpieces—sleek glass boxes designed by the largest architectural firm in the nation, Skidmore, Owings & Merrill, rising up from the clay of the West, announcing the dawn of a new American era.

Don could be a part of that future, just as he'd always hoped. At his previous posting, in northern California, he had worked nights at Stanford to earn a master's degree in political science. Now he was back in Colorado to start a version of the academic life he'd longed for, joining the Academy faculty as an instructor.

The Air Force moved the family into one of a warren of one-story military family houses on the new campus. Theirs was on a hill, with a small patch of grass and a south-facing front door. Don and Mimi set up four bunk beds in the basement level for their eight boys. That worked well until their ninth boy, Matthew, was born in December. Their oldest, Donald, was thirteen now, and he and the brothers close to him in age used the Academy grounds as a playground. They had the run of the place: the indoor and outdoor rec centers, the ice rinks, the swimming pools, the gyms, the bowling alley, even the golf course. No one held them back. In a time of feverish conformity, at the Academy there was also a sense of liberty—the Western frontier spirit, perhaps, or the optimism of a new generation, home from war, building an institution that faced the future with a serene confidence.

Don was like many of the teachers there: World War II veteran hero scholars, young and brash and erudite—and more open-minded than their counterparts at West Point or Annapolis, creating programs in philosophy and ethics that would have seemed out of place at an

older, stuffier military college. His life plan firmly back in place, Don walked the Academy grounds with a certain infectious self-assurance. This was a return of the smooth and seamless Don of his high school president days—and those years in the Navy, playing tennis with the captain of the USS *Juneau*.

There had been a few years in between, it's true, when it didn't seem like things would go Don's way. In his time away from Colorado, Don had hated his assignment in Canada seemingly out of proportion to what the situation seemed to warrant. As a briefing officer, he dealt with classified information, and he spoke in alarmed tones to Mimi about how lax the standards were there; to see papers tossed around without much care seemed to set Don off in a way that Mimi had not seen in him before. His emotional state was fragile enough to force him to take sick leave, first at a hospital at Sampson Air Force Base in New York, and then briefly at Walter Reed Hospital in Washington, D.C. It seemed to Mimi that Don had had an attack of nerves, not unlike what a lot of veterans of the war had, particularly ones like Don who never talked about anything they experienced in battle. But his next posting, in California, was better; Stanford was near his base, allowing him to do graduate work. And now that he was back in Colorado, he, like many of the men in his generation, had come to trust that if you did all the right things in all the right ways, then good things would come to you.

A year before the Academy had even opened, Don had written to the commander in charge of the Academy's organization and construction, General Hubert Harmon, to propose that the Air Force adopt the falcon as its mascot—the same way the Army had its mule and the Navy its goat. Don had not been the only one writing the Air Force to suggest a mascot—the Academy's archives have a folder filled with letters from interested citizens, recommending everything from the Airedale (pun intended) to the peacock—but he was the first to suggest the falcon, something he and Mimi would always point to, once the Air Force took up the idea, as a lasting accomplishment, their contribution to American military history.

The Galvins had brought a few birds along with them on their postings in Canada and California, jamming them in cages in the back of their Dodge woody station wagon everywhere they went. Now that Don finally was part of the Academy, he took over the falconry program, throwing himself into the job like a religious calling. He wrote collectors around the world to build the Academy's stock of birds—accepting two falcons as gifts from King Saud of Saudi Arabia, working to score a few falcons from Japan, and writing the state of Maryland to ask permission to trap falcons there. His cadets flew them in front of tens of thousands of people in stadiums around the country—from Miami University to the Los Angeles Coliseum (in the rain), with a national television appearance during the Cotton Bowl in Dallas in between. He and his birds made it into the pages of *The Denver Post* and *Rocky Mountain News* more than once, and he practically had a standing feature in the Colorado Springs *Gazette*. The birds came home, too. The whole family played a part in training Frederica, a goshawk Don got as his end of a three-way swap with collectors in Germany and Saudi Arabia. When she wasn't scratching the children, Frederica sat on a perch in the front yard, in full view of the neighborhood and well within the sightline of a neighboring Alaskan husky. Once, the bird, which was not tethered, pounced on the dog. The husky ran away with a talon embedded in its fur.

Everybody came to know the Galvins—such a large family, and with their father, the captain who knew everything about falcons. Young Donald became the bird dog for his dad—his "very able assis-

Don and Atholl

tant," Don wrote in *Hawk Chalk,* the newsletter of the North American Falconers Association, which Don had a hand in founding—running ahead and kicking up the rabbits before his father let the birds go. If some failed to return, Donald and a few more of the older boys— John, Jim, and Brian—would wake up at five o'clock to help find them, listening for the bells that had been fastened to their legs for moments like this. From their little house on the hill, the smaller boys sometimes could spot their older brothers and their father through binoculars, climbing up a mountain or rappelling down a cliff.

At home, Don basked in the paterfamilias role while Mimi handled the details. Falconry, again, was helpful to him in this way: Not only did it engage him intellectually, it also allowed him to excuse himself from activities he would just as soon not involve himself in. He had long since taken to referring to the boys as numbers. ("Number Six, come here!" he'd yell to Richard.) When Don started taking night classes at the University of Colorado for a PhD in political science, something had to give. Rather than step away from his duties as the Academy's falconry supervisor, Don gave up the one activity that was centered around the children: coaching his sons' sports teams. He had become, as Mimi described him, "an armchair father."

As the boys grew, their parents' lives only became busier. There was never enough money or time, but the right attitude counted for something, and both he and Mimi continued to believe they had a family others hoped to emulate. Each Galvin boy served as an altar boy. One was responsible for serving a mass each day of the week. Their old friend Father Freudenstein remained in their lives, although he had moved on from Colorado Springs and now served three different parishes out on the prairie. This was not exactly a promotion for Freudy; most priests want to move to larger and larger parishes. But he continued to offer spiritual counsel to Mimi, and he became a favorite of some of the Galvin boys—known for conducting masses in record time, performing his old magic tricks, and showing the older boys the train set and slot machine he kept in the basement of his house, east of Denver. A devoted smoker and unrepentant drinker, Freudy once lost his driver's license, and the oldest son, Donald, when he was in high school, spent a week out on the prairie, staying with Freudy and working as the priest's chauffeur.

In these years, Don saw the boys only insomuch as they were

helping out with the falcons. With Don working or away much of the time, Mimi maintained the home, keeping to a strict routine. She went grocery shopping twice a week, each time bringing home twenty half gallons of milk, five boxes of cereal, and four loaves of bread. More than once, she simply threw out toys that had been left lying around the house. Each morning, she bounced quarters off the boys' beds. Each evening, she made dinner for eleven—iceberg lettuce, cucumbers, carrots, and tomatoes for the salad; minute steaks with a little salt and pepper; a bag of peeled potatoes made into mashers. When he was home, Don would set up four or five chessboards after dinner, line up a few of the boys, and play all of them at once. School nights were for homework and piano practice, not going out. Late at night, Mimi would wash and fold diapers.

In 1959, Don attended a Mardi Gras party in the Crystal Ballroom of Colorado Springs' posh Broadmoor Hotel with a turban on his head and a live, short-winged goshawk in his left hand. He told everyone he was dressed as an ancient seer or mystic. That got his picture in the paper.

Mimi smiled alongside him. She had her own notoriety, thanks to the children. The *Rocky Mountain News* published Mimi's recipe for lamb curry, seasoned with onion and apple and garlic and served with boiled rice and green beans, slivered almonds and artichoke hearts. The headline: SHE SERVES EXOTIC FOODS TO FAMILY OF NINE BOYS.

WHEN HE WASN'T parachuting out of C-47s with the Air Explorer Scouts or studying classical guitar or practicing judo or playing hockey or rappelling down cliffs with his father, the Galvins' oldest son, Donald, was a track star and all-state guard and tackle on the Air Academy High School football team—number 77. Going to his games was often the big family outing of the week. In his senior year, Donald took state in his weight class in wrestling, his team took the state title in football, and he was dating a cheerleader whose father happened to be his father's boss, the Air Force general in charge of the Academy. Donald was, by many measures, his father's son—handsome, athletic, popular—and to his brothers he was a hard act to follow.

He was also, in ways that Don and Mimi either missed or chose to overlook, not precisely what they assumed him to be. Donald was quieter than Don had been in high school, and despite everything he did on the ball fields he was not the sort of person who got elected class president. His grades were average, eventually earning him a spot at Colorado State, not the more selective University of Colorado. And while he looked the part of his father, the carefree charmer, he lacked the charisma to pull it off. From the time he was a teenager, it was as if there was something keeping Donald from connecting with the world in a conventional way. He seemed most at home, at ease with himself, climbing and rappelling from cliffs and raiding aeries in the great outdoors. But whatever sense of mastery Donald demonstrated out in nature didn't play as well around people.

At home, Donald exercised supreme authority over his younger brothers—first as a sort of substitute parent, and then as something less wholesome. When his parents weren't around, he became, by turns, a mischief-maker, a bully, and an instigator of chaos. It would start out innocently enough, before escalating in ways that some of his brothers found terrifying. Don and Mimi would go out—Don training with the falconry cadets, or picking up an extra class to teach at one of the local colleges, or studying for his PhD; Mimi volunteering with the opera—and Donald, the oldest, would have to baby-sit, which he would not want to do. To divert himself, he'd goof around with his brothers: "Open your mouth, and close your eyes, and I'll give you a big surprise," followed by a mouthful of whipped cream.

Then the games would change. Donald would pound his brothers on their arms, right on the muscle, where it hurt the most. And then he would start staging fights. Michael against Richard, Richard against Joseph. He'd have two brothers hold a third one down while he took swings at him, then tell the others to take their own turn on the defenseless, captive brother. The command, to some of his younger brothers, was unforgettable: "If you don't hit him and hit him hard, you're going to be up there next."

At first, all this seemed to happen without Don and Mimi doing much of anything about it. It wasn't that they weren't told—it was that they couldn't believe Donald was capable of the things his brothers accused him of doing. "I begged my parents not to leave him there when I was home," said John—the third son, four years younger than Donald. "Donald, I think, was my father's favorite. He'd take Donald's word over anybody's. In the meantime, I had to go find a place to hide." Mimi also, according to John, "didn't know the half of it. I tried to tell them about my oldest brother, and they just ignored me." *Tattle Tale Tit, your tongue shall be split. . . .*

In Mimi's and Don's view, it might have seemed pointless to get too deeply involved in the gripes and vendettas of teenage brothers. In every family with lots of kids, a pecking order was inevitable. Donald would take command when Don and Mimi weren't around, and Jim would seize power when Donald was gone. "The older brother would control the situation," remembered Michael—the fifth son, who was eight years younger than Donald—except one time when Michael dislocated his elbow in a fight with Richard, who was younger, and Richard rose in triumph. Michael, in turn, once slapped Mark so hard, half his face turned purple with a bruise. The walk to school wasn't safe. If you didn't form a new alliance each day with a few brothers, you were basically asking to be dominated.

These conflicts, Don and Mimi sometimes thought, were best settled between the boys. Interceding too much might send the wrong message, and the boys might never learn to get along with one another on their own. And even if they had wanted to determine who was wrong in every instance, they would have had trouble figuring out whose fault it was. Because while Donald might have ruled over the other boys with an iron fist, Jim never stopped gunning for Donald's position at the top of the heap.

IF DONALD WAS the model son, Jim was more of a maverick. This meant embracing the James Dean and Marlon Brando spirit of the time—the leather jacket, the fast car, the defiant snarl. He had tried being more like Donald first and came up short. As an end receiver and defensive back, Jim had been good enough to, one time, block a punt and score a touchdown in the same play, but he never could outperform Donald on the football field—or in falconry, for that matter. Soon, he saw no upside in trying. Jim couldn't fail to notice his parents' expectations and attention always swerving past him and toward his older brother, and that angered and shamed him. There was, naturally, one person onto whom he could direct all that rage. Which was how, from his teenage years onward, Jim always seemed to have a score to settle with Donald. "It was like a pact with himself," Michael said.

Jim and Donald never seemed to stop wrestling—in the basement, in the bedrooms, on the shrubs in the backyard. Jim was smaller, and so when Donald beat him, he'd go off and lift weights, or try to round up some of his younger brothers to gang up on Donald. That never worked. The other brothers were afraid of them both. Once, Jim slammed a storm door in Brian's face, cutting his mouth. The skirmishes extended beyond the nights Mimi and Don were out, and into the daytime when they weren't around, and then every waking hour. Once the fights started spilling into the living room, Don and Mimi knew they had to intercede. Michael remembered being in third grade and seeing his father, usually so aloof, running at teenage Donald at full speed and tackling him, to keep the boy from hurting one of his other brothers. This struck Michael as impressive at the time. But Don must have known he was on borrowed time. Donald was a football star. All the boys were only growing in one direction.

Different command styles are appropriate in different situations, and so Don searched for the right approach. At first, to him, this seemed like an issue of setting the right tone. Something about the family atmosphere seemed too highly charged with young male vigor, and it was his job to help them find their way, each of them, toward manhood. Ever the benevolent patriarch, Don would try to sell the boys on books that could improve their personalities, smooth out the rough edges. *The Power of Positive Thinking* was one. Another was *Psycho-Cybernetics,* Maxwell Maltz's popular 1960 self-help book, which introduced much of the public to the idea of creative visualizations. Don thought these books might offer the boys a road map out of their conflicts. He'd get them all around the long dining table and lecture about harmony. When that didn't work, Don decided that he could at least impose order, something the military had taught him well. So he brought home boxing gloves. And he made a new rule: No fighting without them.

Richard—the sixth son, nine years younger than Donald—remembered the dread he felt when he put on those boxing gloves. "All the brothers were all-state athletes, you know, in top-notch shape," he said. "So when a fight broke out, it was a real fight."

THE GALVIN HOME became a place where two different realities existed at the same time: the wrestling pit and the church choir; the wildness of the boys and the model family Don and Mimi believed they had. A little mischief could always be waved off, especially on a military base where competition and power and might were almost part of the drinking water.

But for many brothers—John, Michael, Richard, and Matt—there was a growing sense of being lost in the shuffle, even neglected, feeling less than safe, treated like a number and not a person, raised to take the illusion of protection as the real thing.

The private and public faces of the family were sometimes hard for others to reconcile, too. Visiting their cousins in Queens, the boys seized every unsupervised moment to break every rule in the house—climbing up on top of the garage roof, shooting at windows and birds with a BB gun. The East Coast cousins were both thrilled and scandalized. Then, months later, the cousins would get Christmas cards of Don and Mimi and the kids, a saintly family tableau, everyone dressed perfectly in pajamas around a tree. The disconnect seemed strange to them, even then.

For the moment, Don and Mimi chose not to see what was happening as anything other than roughhousing. These were boys, a lot of them, living in close quarters. It would be unrealistic to think that they would never fight. And the oldest, Donald, was still such a source of pride—a clean-cut teenager who was featured in a photo taking up nearly half the front page of *The Denver Post,* rappelling from a falcon-nesting spot high up on Cathedral Rock. Just like his father.

PETER, THE TENTH son, was born on November 15, 1960. This time, Mimi had a long stay in the hospital afterward with a severe prolapse, along with a blood clot in her left leg. Now there were fewer jokes about how Mimi ought to wear garlic to bed to fend off her husband at night. Her doctor gravely told her that her childbearing years were behind her. Fifteen years of more or less continuous pregnancy, labor, and delivery would seem to be enough for anyone. But Mimi did not seem interested in listening, even when others pleaded with her.

"Really, dear, you should give poor 'Major Galvin' a turn at the hospital," Mimi's paternal grandfather, Lindsey Blayney, wrote to her. "But, seriously, I am concerned about you."

In truth, asking Mimi and Don to stop at ten children was like asking a marathoner to stop at twenty-five miles. Stopping struck them as ludicrous—not now that Don was rising in status at the Academy, and they both were feeling comfortable back in Colorado. Besides, as if it weren't obvious to everyone yet, they still did not have a girl.

In 1961, mere months after giving birth to Peter, Mimi became pregnant for the eleventh time. Over Christmas, shortly before she was due, she and Don and the ten boys assembled for a photo on the grand staircase in Arnold Hall, the central gathering place at the Academy. The boys wore identical Eton suits from Lord & Taylor, paid for by Mimi's mother, Billy. Mimi's grandfather Lindsey called the photo "startling" when he received his copy in the mail—all of those children, with no sign of stopping. He predicted that Mimi's next delivery would be twins, which would make for an even dozen.

He was wrong, but Mimi did break new ground on February 25, 1962, by delivering Margaret Elizabeth Galvin. *The Greeley Daily Tribune* of Colorado broke the news with a short article: AT LONG LAST, IT'S A GIRL! Don had predicted as much, but that didn't keep him from joking to his cadets, "Dang, that was supposed to be my quarterback."

Mimi didn't bother disguising her joy. "She is the most beautiful one yet to come into our family," she wrote in one letter. "She makes each day a mother's dream." She also found room to boast about the rest of her children. Donald, she wrote, "plays a lovely classical guitar and is an outstanding high school athlete. His grades are wanting but as his principal states, 'I wish all the boys were as fine as Don.'" Jim, Mimi said, was "a good all around boy and a great help to me." John (number three), with "curly brown hair and sparkling blue eyes," played clarinet and piano with dedication, plus ran a paper route with sixty-five stops. Brian (number four) was "our shining prodigy at the moment," making Chopin's tragic overture "most tragic" and Jacques Offenbach's "Can-Can" from *Gaîté Parisienne* "most gay." Michael (number five), who played the French horn and liked to read, was "the delicate member of the family." Richard (number six) was a "mathematician" who also wanted private piano lessons, "but with two taking private lessons he will have to wait a while." Joe (number seven) was in kindergarten, learning his letters and numbers and phonetics. And Mark, Matthew, and Peter (numbers eight, nine, and ten) "are my constant companions at home. Like teddy bears, they are always getting into one thing or another. One day I found they had vacuumed the dish waste out of the sink with a new Electrolux!"

Money was going to be tighter than ever, no matter how many extra political science classes Don could teach at the local colleges. Parochial school clothing—two pairs of shoes, two shirts, and pants—cost about a hundred dollars per child. Between feedings for the baby and snacks and meals for the others, Mimi put her Bernina sewing machine into overdrive, making all the clothing herself. But Mimi would say the horizon turned pink on the day Margaret was born. As if by magic, the world had finally cooperated with her and given her what she'd wanted the most. She also said she wanted another—a twelfth—which delighted Don, but alarmed her obstetrician.

When their twelfth and final child, Mary, was born on October 5, 1965, Mimi was forty years old. Her doctor told her flat-out that if she got pregnant again, he would refuse to treat her. He urged her to have a hysterectomy, and Mimi reluctantly agreed. She and Don figured there would be grandchildren to look after, sooner or later.

By the end of November, Mimi was back on her feet, announcing in the Colorado Springs *Gazette* that roles had been assigned for the Colorado Springs Opera Association's upcoming production of Verdi's *The Masked Ball*. That same year, the local chapter of Notre Dame's Knute Rockne Club of America named Don Galvin its Father of the Year. Mimi had to laugh. "I had all the babies," she would say, in that mixture of sweetness and sharpness she'd long since perfected. "He got all the degrees and all the applause."

WHEN HE WAS about seventeen, Donald smashed ten dishes to pieces one night—all at once, while standing in front of the kitchen sink.

Don wrote it off. So did Mimi. Donald was a teenager, moody. It was the 1960s. Other kids were doing worse.

But Donald knew there was something wrong. He'd known for a while.

He knew that despite the similar hairline and strong jaw and athletic talent, he was not like his father, and that he was never going to be. His grades were mediocre, not the grades of the son of a man whose children considered him the smartest man in the world. His

fights with his younger brothers were little more than his own ham-
fisted attempts to control them the way a father ought to. He failed
at that, too.

He knew that being a star on the football field and having a
friendship with another person were two very different things. Sometimes, he would say later, he thought of people as kind of like IBM
cards he sorted through his own computer for information he could
use. He knew that made him unusual.

Donald recognized how trapped he often felt—frustrated that he
was not the person he wanted to be. But at other moments, increasingly often, he seemed completely oblivious—a stranger to his own
motivations and actions.

Something was happening, and he couldn't figure out what. More
than anything else, he was afraid.

DON
MIMI

DONALD
JIM
JOHN
BRIAN
MICHAEL
RICHARD
JOE
MARK
MATT
PETER
MARGARET
MARY

CHAPTER 6

In the autumn of 1963, the Galvins moved from their quarters on the Academy grounds and into a newly constructed split-level house in Woodmen Valley, a densely pined collection of dairy farms a few miles out from the center of Colorado Springs. Don had paid a few thousand dollars for three acres of land along Hidden Valley Road, a four-mile dirt trail that terminated in a gravelly cul-de-sac. Theirs was one of the first in a new line of suburban homes meant to cater to Academy families who wanted a little more room. Before construction, Mimi tied ropes around every tree and bush on the property to make sure the contractors wouldn't cut them down.

To many of their Academy friends, Woodmen Valley was backcountry, the middle of nowhere. But Mimi, whose feelings about the outdoors had reversed completely since first coming to Colorado twelve years earlier, loved how unspoiled it felt. So much of Colorado Springs had been built up and paved over for the military; not just the Air Force Academy, but Peterson Air Field, Fort Carson, and most recently NORAD, the nuclear defense coordination headquarters embedded in the Cheyenne Mountain defense bunker between Colorado Springs and Pueblo. Woodmen Valley was just a fifteen-minute drive from the center of downtown Colorado Springs, and yet living there, to Mimi, felt as far away from the nuclear age as could be—more timeless, more natural, more authentic.

A short walk from their new house stood a convent that had once been a tuberculosis hospital—the Modern Woodmen of America Sanatorium, for which the surrounding Woodmen Valley had been named. The valley's geology was a little less red and more white than in the rest of Colorado Springs—leftover feldspar and quartz gravel from the eroding mountains that had settled there, millions of years earlier. Beyond the pines, there were enough large rock formations to have once sustained a tourist attraction called Monument Park. The boys could fill their days exploring the rocks famous enough to be named: Dog Rock, Grandma Grundy, Anvil Rock, the Dutch Wedding Rocks. But the magic of Hidden Valley Road was that it had enough trees and rolling hills to seem like a forest, tucked away from the avalanche of rock. Deer wandered by the patio door at breakfast, and blue jays squawked from the branches of the pine trees overhead.

The house itself was an early-1960s low-slung box, more long than tall, coated with the usual mix of siding and stone. Inside, a carpeted sitting room connected to a kitchen and dining room, just large enough for a mammoth dining table made by a family friend that could fit two people on each end, if needed, and six down each side. From the front hall, one half staircase led up to the bedrooms while another led down to a sublevel that, by necessity, the Galvins used for more sleeping space. Margaret, Don and Mimi's eleventh child, was eighteen months old; she would share a room with Peter, upstairs near her parents, with pale lime-green carpet and a large pine tree outside the window. Mark and Joe shared a room on that floor. On the ground level, Peter and Matt shared a room with twin beds. The older boys still left at home were in the lower level, sleeping on corner couch units that unfolded into beds at night.

For Mimi, everything about the house on Hidden Valley Road shouted *just enough:* Enough of a living room to accommodate wrestling matches, enough of a kitchen to cook all day for the family, enough space outside to breathe when you needed to, or to play football, or ride bikes, or fly falcons. Mimi and the older boys put three coats of paint on every wall. And she herself started work on a rock garden in the back, near where Frederica the goshawk stood guard. Don built a large A-frame mews on top of the hill in the backyard that

Mimi and all twelve children, standing in age order from Margaret to Donald, with Mary in Mimi's arms. Photo by Don.

housed more birds, including Hansel and Gretel, two hawks they'd take flying on the sprawling lawns of the Carlson family's nearby dairy farm. Their prized birds—Frederica, followed by her successor, Atholl—also were permitted to perch on the living room coffee table. For the first time, perhaps, the Galvins were home.

DONALD'S FRESHMAN YEAR at Colorado State coincided with the move to Hidden Valley Road. None of his inner fears were on display, not to his family. He told them he wanted to be a doctor, and he saw how proud they were to hear that. His job, after he left home, was to maintain that veneer. On Hidden Valley Road, the power vacuum among the brothers was seized by Jim.

Having long since stopped competing on his older brother's playing field, Jim instead set out to dominate in the areas Donald seemed weakest. If Donald won childhood, the first round, Jim would win the next round, real life. Jim tried assuming the role of the younger boys' cool big brother—the brother in the biker jacket, the brother who drove a black '57 Chevy, the brother most likely to offer to sneak

a little Bacardi into your Coke. The younger Galvin brothers appreciated it sometimes, but remained mostly wary, particularly after Jim
started hitting on any girls they brought home with them. Jim liked
being known as provocative, and if he came off as menacing, so much
the better. He had, he thought, a certain confidence, or brashness,
his other brothers lacked. "At sixteen, we knew something was wrong
with Jim," said Richard, who was seven years younger, "but we just
thought it was okay, just being a boy—out drinking, carousing, de-
linquent activity, skipping school."

No longer burdened by the Galvin family's demands of perfec-
tion, Jim drank more than Donald had, went out more, and got in
trouble more—culminating in a stunt that got him kicked out of Air
Academy High School in the middle of his senior year, the same year
that the family moved to Hidden Valley Road. He and a friend were
at the Academy's jet center, clowning around on one of the planes.
Jim was inside the cockpit, and his friend was just outside, when Jim
pushed a button that made the plane move slightly, enough to send
the other boy flying backward, colliding with the tail of the jet. An
inch or two in a different direction and that boy might have died. Jim
was forced to transfer to the local Catholic school, St. Mary's. This
would have been a shock if it had happened to Donald. Not so with
Jim. The consolation of being a washout is the benefit of low expectations. Jim had nowhere to go but up.

Nor did the expulsion humiliate the Galvins as it might have another proud, striving family. Mimi understood how to take the worst
possible bad news and cast it aside, moving on as if such a thing had
hardly mattered to begin with. She'd watched her own mother do it,
when her father left the family in scandal. Don, too, knew how to
blot out the darker aspects of his life, leaving a number of subjects
undiscussed: the horrors he saw firsthand during the war, his failure
to advance in the Navy, his troubling hospitalization during his Air
Force posting in Canada. And so now that they had hit their stride in
Colorado, they were not about to let their strong-willed son's ridiculous mistake define them. It was simple enough for Don and Mimi
to decide that the problem of Jim, as they saw it, was already on its
way to being solved. He was finishing high school and would soon be
off on his own. Maybe he would take a year of community college to
clean up his academics for a real four-year program. But no matter

what, as Don always told Mimi, Jim would have to grow up sooner or later. All the boys would.

For Mimi, coming to Hidden Valley Road was meant to signal the beginning of her family's long-awaited von Trapp family–style idyll. With Donald no longer living at home, and Jim about to leave, she felt as if they all had that perfect life now, almost within reach. What if what they'd needed all along—Donald, Jim, all of them— was some extra space to spread their wings? She wanted a house filled with music, and she enlisted the boys to help her. The boys learned piano on a bargain-basement $850 baby grand that Don and Mimi found in a shop downtown. John and Brian and Matt and eventually little Peter all played flute. On weekends, Mimi would throw a symphony on the record player and tell the story behind it, explaining the music with encyclopedic detail. When the boys got a tape recorder, they'd tape the Saturday morning Metropolitan Opera broadcast for her, and Mimi would play it all week long, alternating it with sing- alongs to ballads and folk songs by Burl Ives and John Jacob Niles. In the neighborhood, the Galvin kids played kick the can and cap- ture the flag and kickball and Simon Says with the neighborhood children—the Skarkes, the Hollisters, the Turleys, the Warringtons, the Woods, the Olsons. In the fields and forests of Woodmen Valley, Mimi taught the children to identify wild animals, like the bobcat that lived in the small dark cave in the white cliffs down the road.

As the 1960s progressed, the habits and motivations of the younger generation became more mysterious and frightening to the parents of the Galvin children's friends. Not Don and Mimi. The Galvins remained good Catholic New Frontier–era liberals, permis- sive socially but disciplined domestically, tolerant in their hearts yet strict in their ways. They prayed for the president who had died just a few weeks after their move to Hidden Valley Road, and they prayed for the president who had taken his place, and as the conflict in Viet- nam escalated, Don, a colonel in the Air Force, held his tongue about how he felt. Only later would he tell his boys that the unfortunate ones who had been sent to fight in Southeast Asia were nothing more than, in his words, "assassins in uniform." Most of his sons went to parties, played rock 'n' roll music, and stayed out late. As long as they came to mass on Sundays, dressed appropriately, all was as it should be.

THE GALVINS HAD done all the right things in all the right ways, and now, just as Don had always trusted they would, good things seemed to be coming their way.

Just before the move, Don, who was nearing the twenty-year mark of his military service, transferred to a new post at NORAD. His title was information staff officer. This was another job briefing generals, like the one he'd had years earlier, only this time, the job had an element of public relations, sending him out to deliver speeches to clubs and organizations around the country, explaining the international defense control center that coordinated the continent's first ballistic-missile early warning system and the deployment, if or when the time came, of nuclear weapons located at eight hundred separate military installations in the United States and Canada. Back home, the Galvin boys, who along with their schoolmates were the first generation to grow up living with the prospect of possible nuclear annihilation, thrilled at eavesdropping on their father after dinner, when he filled in generals with end-of-day briefings on the kitchen phone. Back at headquarters, Don gave tours to reporters and visiting public officials, often working in mentions of his bevy of children and his beloved Academy falcons. Colonel Galvin "was apparently gone on birds," wrote a columnist from the *Daily Star* in Hammond, Louisiana. "He kept telling the group how he trains falcons (to hunt) and was instrumental in getting the Academy's sports teams named the 'Falcons.'"

The greatest of all the good things happened in 1966, when Don retired from the Air Force and started a new career as a grant-in-aid man, overseeing programs funded by the federal government for the benefit of the states—first as the vice chairman of the Colorado State Council on the Arts and Humanities, and then as the first full-time executive director of the Federation of Rocky Mountain States. This new organization counted seven states in the American West as members, from Montana down to New Mexico. Soon enough, Arizona would make it eight. The Federation was a quasi-governmental group, formed to help the region attract industry, banking, the arts, and major transportation projects. The governors of each of the member states took turns heading the group. But the real man in charge, day to day, was Don Galvin. He was putting both his political science degree and his military experience into action as a sort of domestic diplomat—a liaison between the government and the private sector and nonprofit worlds. The older boys who were still at home were in awe of him. "He was telling governors what to do," said Richard, the sixth son, who was twelve when Don started the job. "You knew he had the presence, but man, when you heard his voice, it resounded."

With his new career, Don's—and, by extension, Mimi's—horizons were only broadening. What was once a quiet life in Colorado among the falcons now seemed like a stepping-stone to the world stage. In Washington, Don lobbied for a new railroad from Albuquerque, New Mexico, to Cheyenne, Wyoming; and a pipeline to bring water south from Canada or Alaska; and the western United States' first public television station. The Federation pooled risk capital for experimental industrial projects, worked to find new mineral and water resources, formed a science advisory council for technological development, and promoted tourism with touring art exhibits and support for the Denver, Phoenix, and Utah Symphonies and the Utah Civic Ballet, which Don renamed Ballet West. The new name actually was Mimi's idea: "*Utah Civic* spells Mormon all over it," she'd said with a roll of her eyes. But Howard Hughes had just named his new airline Air West; maybe if they followed his lead, Mimi suggested, Hughes would donate one day?

With money from the National Endowment for the Arts, Don started offering residencies to the East Coast's most prestigious and accomplished dancers and choreographers and conductors. By the late 1960s, Don and Mimi and whichever children were too small to leave at home on their own would travel to Aspen and Santa Fe for concerts, fund-raisers, conferences, and galas. Which was how, with the Federation, Mimi's old dreams of a life of art and culture and the best of everything really were coming true—first the dream house, then the dream life.

In Santa Fe, the Galvins were regulars at parties where the guest list often included Georgia O'Keeffe—in her signature black hat and long black skirt, her hair in a long braid down the middle of her back—and Henriette Wyeth, Andrew's sister, who demanded to paint Don and Mimi's little girls, Margaret and Mary, in their gossamer organdy dresses that made them look like they'd stepped right out of a double portrait by Gainsborough. For Mimi, very little could match the thrill of visiting Henriette Wyeth's ranch in Roswell, New Mexico, standing in the barn where she and her husband, the artist Peter Hurd, painted, and seeing Hurd take her two little girls on a hike to look at the orange trees and the sagebrush that made little Margaret sneeze. Or having breakfast with the legendary conductor Maurice Abravanel and choreographer Agnes de Mille (who, like Georgia O'Keeffe, showed extraordinarily little interest in young Margaret and Mary). Or watching Don as he sweet-talked David Rockefeller into funding the Federation's new public television project.

They made new friends, too, like the oil wildcatter Samuel Gary, whose 1967 strike in Bell Creek Field in Montana tapped an estimated 240 million barrels of oil—the largest oil strike west of the Mississippi at that time. Sam relied on Don and the Federation for help in building out Bell Creek into a town that could support hundreds of new oil workers. If the main drag of Bell Creek needed a new traffic light, Don Galvin was a phone call away. Through the late 1960s, with Margaret and Mary in tow, the Galvins visited with the Garys at their house in the refined Cherry Hills section of Denver. Sam and his wife, Nancy, had eight children, and a few of the girls were close in age to Margaret and Mary. The children would play together while the grown-ups would play tennis or talk politics. The Garys loved watching Don with his falcons; Don's fame as the Air Force Academy's falcon man preceded him. Once, in Colorado Springs, Don and Mimi enlisted young Donald to teach Sam and Nancy and some of their children how to rappel off the cliff at Cathedral Rock. Another time, when the Garys flew Don and Mimi to *Swan Lake* in Cedar Springs, Idaho, in their tiny, unpressurized private plane, Mimi got dizzy during the flight and passed out.

Back home, Mimi and Don became regular guests at dinner parties, where Don held forth with authority on politics and industry and the arts. All eyes were on her accomplished husband. Mimi felt she had it all on those nights. Don was handsome, intelligent, and a little flirtatious. Her friends would call him Romeo.

NOTHING IS FREE, and before long, Mimi put her finger on the price. More than Don, she saw how her nose was pressed up against the windows of this world. She had no college education, and she and Don had no wealth. Her own pedigree, Grandfather Kenyon and his levees, mattered very little among the millionaires of the new

West. At best, they were the help. Even at their most benign, Sam and Nancy Gary, their new multimillionaire friends, were living reminders that the world that Mimi and Don were traveling in—the world of the Federation and governors and oil wildcatters and world-class artists and dancers and celebrity orchestra conductors—was not really their world at all.

And, of course, their world was not as perfect as Mimi had wanted. She would not have admitted this to herself at the time, much less told another soul about it. But if she needed reminding, she only had to wait for visits home from her oldest two boys. Donald and Jim continued to fight, with each other and with their younger brothers. Every visit to Hidden Valley Road—Christmas, Easter, Thanksgiving, Christmas again—ended in bruises. Richard remembered once watching as Donald ran down the road after Jim, caught up to him, and knocked him to the ground with an uppercut. He had never seen anybody punch someone so hard in his life.

Mimi had surprised herself by being relieved that her two oldest boys were out of the house, on the pretense that Donald and Jim were, in theory, nearly adults and capable of making their own decisions. Each time they came home put the lie to all that. But she also was aware that the slightest acknowledgment that all was not well in her family risked coloring everything else about her life—Don's new professional prospects, the standing of the other children, the reputation of them all.

And so Mimi tended to agree, most of the time, when her husband said what he'd always said when there was something wrong with the children: that the boys should not be coddled; that they should leave the nest, make their own mistakes and learn from them, take responsibility for their actions, grow up.

And she thought about how perfect their life was otherwise. And how fragile her husband's happiness had always seemed to her. And how sometimes it seemed as if the slightest move in any direction could bring the whole place toppling down.

CHAPTER 7

On September 11, 1964, Donald Galvin, at the start of his sophomore year at Colorado State in Fort Collins, paid his first visit to the campus health center. He had come in to be treated for a minor injury to his left thumb, a bite mark from a cat. He offered no explanation for what had happened—no reason why the cat would have felt so provoked that he'd bite and not just scratch.

The next spring, Donald returned to the health center. This time, his problem was more personal, yet every bit as peculiar. He said that he'd learned that his roommate had caught syphilis, and that he was afraid that he might catch it from him by accident. Donald, who had told his parents that he wanted to study medicine one day, had to be disabused of the notion that he could get the disease in a way other than sexual intercourse.

A few weeks later, in April 1965, Donald visited the health center for a third time. He said he was at home, his family's place on Hidden Valley Road, when one of his brothers, he did not say which one, got the jump on him, attacking him from behind. Diagnosed with back strain, he spent the night in the infirmary.

Then came the fire.

One night in the fall of 1965, Donald staggered through the

health center doors with burns on his body. His sweater had caught fire, he said, during a pep rally. After a little back-and-forth, it came out that Donald had jumped straight into a bonfire. Maybe he did it to get attention, or to impress a friend, or as a cry for help. He could not say.

THE STAFF PULLED Donald out of his classes and sent him for a psychiatric evaluation. Major Reed Larsen, a clinical psychologist for the Air Force Academy Hospital, saw Donald four times over the next two months. This was the first time that a mental health professional examined Donald, and the first time that Donald's parents were forced to face the possibility that all was not right with their oldest son. But whatever fears Don and Mimi had about Donald subsided when Major Larsen came back with his report. "Our findings showed no evidence of a serious thinking disorder, nor of symptoms secondary to a psychotic process," he wrote on January 5, 1966.

Don and Mimi were reassured, even if the endorsement was hardly full-throated. To begin with, the major noted that one of Donald's sessions took place with the assistance of sodium amytal, one variety of truth serum. Amytal interviews in psychotherapeutic settings weren't entirely unheard of, but they were usually saved for patients who are having difficulty communicating—and, perhaps, exhibiting the signs of the catatonic variety of schizophrenia. Even so, the major recommended that Donald be allowed back to school, provided he continued to receive psychiatric help. "We did discover a number of emotional conflicts which, I feel, are disturbing enough to Mr. Galvin to account for his erratic behavior while at school," he wrote. Such treatment could be paid for, he said, by the military's new Medicare program for dependents.

What was bothering Donald so much that he ran into a raging fire? Before anyone could find an answer, he propelled himself back into campus life at the start of 1966, determined to make up for lost time. Donald desperately wanted to connect with people now, especially females, even as he seemed rather naive about how to find a girlfriend. The distance from others that he'd been feeling seemed even more pronounced. But he was still athletic and handsome, and he hoped there was still every chance that he could become the man his parents thought he could be.

He started seeing someone, a classmate named Marilee. Within a few months, they were even talking about marriage. This seemed fast—but not if, like Donald, you were eager to lead a normal life, to have sex without it being considered a sin, to have a family like his own family, to be all right. But the family never got a chance to get to know Marilee. When the couple broke up, Donald was shattered, and he kept the news to himself as he scrambled to make things right. On the phone with Marilee afterward, he racked up $150 in long-distance charges. He couldn't pay his rent, but he also couldn't bear to admit that to his parents. Donald's solution was to search for a place where he could live for free—a place to hide while he figured out what to do next.

In the fall of 1966, Donald found an old, abandoned fruit cellar near the campus—a room with electricity and an old heater, but no water. He slept on a mattress there alone, not sure of how he might climb out of the hole he'd dug for himself. Days turned into weeks, then months—until, on November 17, Donald returned to the health center, reporting, once again, that he'd been bitten by a cat.

When the doctors learned that this was his second cat bite in two years, they sent him that same day for a full work-up with a psychiatrist. It was there, finally, that the extent of Donald's troubles became clear. He seemed to open up to these doctors in a way he hadn't before, perhaps to anyone else. The intake notes mention more "bizarre self-destructive things" Donald said that he had done: "Has run through bonfire, put cord around his neck, turned on gas, and even gone to a funeral home to price caskets—all of which he cannot give adequate motivation for."

A noose, a gas switch, a funeral home. Donald was fixating on death, on ending his life. This disconnection he'd always felt wasn't going away at college—it was getting worse, manifesting itself in new and frightening ways.

While under observation, Donald's free fall continued. He told one doctor that he had a notion that he had murdered a professor. Days later, he shared another fantasy—this one about killing another person at a football game. He also talked more about his past, including a new admission that the doctors found especially troubling. The hospital notes were brief: *2 suicide attempts at age 12.*

Exactly what those attempts amounted to, no one could say. There was no telling if Donald had ever told anyone else about them—or, assuming they did happen, that his parents had ever known. But the doctor treating Donald had heard enough. Especially after learning what had really happened with the cat.

"He killed a cat slowly and painfully," the doctor wrote in his notes. "The cat had been living with him for two days, and apparently brought in another cat (probably male) that made the place smelly. The cat scratched him. Doesn't know why he killed the cat nor why he tormented. Got emotionally upset as he discussed the behavior."

Donald was more than baffled as he was relating this. He was frightened.

"This boy represents some risk to himself and possibly to others," the doctor wrote. "Possible schizophrenic reaction."

IN THE CAR, Donald muttered about God and Marilee and some people from the CIA who were looking for him. Back home, in the kitchen, Donald exploded in a panic—shrieking "Get down! They are shooting at us!" Everyone around him jerked around to see if what he was saying was true.

It was the end of 1966, just as Don had started his new job with the Federation of Rocky Mountain States—the new life for them all, about to begin. The doctor at Colorado State said it would be impossible for Donald to continue in college until he received more evaluation and treatment. Don and Mimi drove to Fort Collins at once to check on their son. When they found him, Donald was washing his hair with beer. They decided to take him home. But now that he was there, they did not have the slightest idea what to do with him.

Donald needed help. But what help was available to him? Assuming he'd be willing to go, a private facility like Chestnut Lodge in Maryland or the Menninger Clinic in Topeka—or, closer to home in Colorado Springs, a hospital called Cedar Springs—was too expensive an option for the Galvins. The public hospitals, meanwhile, were a terrifying prospect, places where the peace was kept using neuroleptic drugs and restraints—the stuff of Samuel Fuller's nightmarish film *Shock Corridor,* released in 1963. In 1967, the state of Massachusetts made headlines by litigating to stop the distribution of documentarian Frederick Wiseman's film *Titicut Follies,* an exposé of the inhuman conditions at that state's Bridgewater State Hospital, filled with images of inmates stripped naked, force-fed, and bullied by the people who were supposed to be keeping them safe. In Colorado, the very large state mental hospital in Pueblo, about an hour's drive from Hidden Valley Road, was best known for treating schizophrenia with insulin shock therapy and a powerful drug called Thorazine. Don and Mimi would have to exhaust virtually every other option on the table before agreeing to send Donald to a place like that. A state hospital like Pueblo was for hopeless cases, not healthy young men like their son.

There was an alternative to the brutal public institutions, but that alternative also was hardly attractive to Mimi. The psychoanalytic approach advocated by Frieda Fromm-Reichmann and others held sway at the Colorado Psychiatric Hospital in Denver, part of the university system. This hospital was steeped in the teaching of schizophrenia as a psychosocial disorder, focusing on the "psychodynamic" origins of mental illness—the schizophrenogenic mother. Mimi and Don may not have known the particulars of this approach—how a psychoanalyst would want to know exactly how Donald was raised, and if there was something they could have done differently—but they understood the threshold they would be crossing by sending their son to a mental hospital of any kind.

Again, they thought, were things really so far gone? After all, it seemed clear that diagnosing schizophrenia was—and in many ways remains—more of an art than a science. None of the symptoms, taken by themselves, were specifically characteristic of the illness, and so doctors could only diagnose it by excluding other possibilities. The American Psychiatric Association had published the first edition of the *Diagnostic and Statistical Manual of Mental Disorders,* or DSM, fourteen years earlier. The definition of schizophrenia was about three pages long, and included the subtypes originally proposed by Eugen Bleuler—hebephrenic, catatonic, paranoid, and simple schizophrenia—and added five more: schizoaffective, childhood, residual, chronic undifferentiated, and acute undifferentiated. The definition was roundly panned: In 1956, one prominent psychiatrist, Ivan Bennett, called the DSM's definition of schizophrenia "a wastebasket diagnostic classification," preferring instead to focus on what drugs might be helpful in treating the symptoms. Since then, the DSM has changed its description of schizophrenia with each successive edition, often tailoring it to the prevailing style of treatment. The second edition of the DSM, published in 1968, added "acute schizophrenia," characterized by hallucinations and delusions and nothing else. But there would continue to be no consensus on what schizophrenia actually was. A single illness, or a syndrome? Inherited, or acquired through trauma? Don and Mimi understood that for people in their son's shoes, whether you even had schizophrenia or not often depended on the priorities of the institution where you were being examined.

There was no talk of prevention. There was very little discussion of a cure. But one thing seemed true: If they admitted Donald to anything resembling a mental hospital, the only certainties were shame and disgrace, and the end of Donald's college education, and the tainting of Don's career, and a stain on the family's position in the community, and finally the end of the chance for their other eleven children to have respectable, normal lives.

Which was why, for Mimi and Don, the most sensible—or at least the most realistic—decision was to hope, somehow, that things would get better on their own. The more they thought about it, the more they decided to be optimistic. Why *couldn't* he move on from Marilee, find his footing again, move out of that fruit cellar and into the dorms, and get better? They needed to believe that he could. And so they searched for someone they knew and trusted to treat Donald—who could help him through this crisis, get him back to college, put him back on track.

Their obvious first stop, they thought, was back to the hospital at the Air Force Academy, where the Galvin family was well known, and where they hoped to be able to help guide the process to a good outcome. This time, Donald was examined by Major Lawrence Smith, a physician who knew the Galvins well. He had been at the Academy since 1960, overlapping with Don for three years, and he had followed young Donald's football career.

On December 8, Major Smith wrote a letter to Colorado State University on Donald's behalf, blaming what he called Donald's "acute situational maladjustment" on a freak confluence of bad breaks: his substandard housing situation, his breakup with his girlfriend, and the stress of final examinations. The tone of the major's letter was generous and reassuring, filled with goodwill. "I agree that his reaction in December when he saw you was quite bizarre," he wrote. "However, I feel that he has recovered from the incident, has insight into the situation, and to the best of my knowledge will probably not repeat this behavior."

For a second time in the space of a year, Don and Mimi had secured a scandal-free return to college for their son. The major did not mention Donald's killing of the cat, or his homicidal fantasies. There was a good reason for this: Major Smith hadn't been told about any of that. He had never spoken with anyone who had examined Donald at Colorado State. They'd never had the chance to let him know.

Donald, naturally, didn't volunteer it.

DONALD RETURNED TO Colorado State just after Christmas break. The fruit cellar was a thing of the past. He was out of isolation and back in the world of his classmates again. He kept seeing therapists at the health center, sitting for occasional psychiatric evaluations. After one, his evaluator wrote, "This student is not psychotic."

Once again, he seemed in a hurry to be all right, to be the son his parents wanted. He was even dating. That spring, he announced that he had met someone new, a successor to his old girlfriend, Marilee. Her name was Jean, and she was tall and broad-shouldered—a tomboy, as Donald once described her. Physically, Jean was a good match for Donald, who was still built like a football player. Like Donald, she was ambitious. She wanted to get a PhD, and Donald still was hoping to become a doctor.

They were together for several months before Donald told his parents that, once again, he was engaged. Mimi and Don were torn. In some small way, they took it as a positive sign that Donald wanted to get started on the rest of his life. They even granted Donald a certain degree of credit for having enough forethought to plan a marriage without a pregnancy forcing the issue. They also knew, from personal experience, that in a situation like this, when you're young and determined, the objections of your family don't mean a thing. And Mimi also was, in at least one respect, a little relieved. She and Don had been keeping Donald's breakdowns secret from the world, hoping that perhaps they could be forgotten. She wanted nothing more than for Donald to right himself. How could she be opposed to the idea—the hope—that Donald might settle down, find direction in life, become predictable, grounded, successful, even happy? Wasn't this how the story was supposed to go? Boys and girls met and fell in love and got married.

But of course they knew that marriage was a terrible idea. Everyone did. Even beyond his personal problems, the match seemed off for at least one very important reason. Those who knew them warned Donald that Jean was very clear about not wanting children. She wanted to pursue graduate work in genetics and help cure diseases. Children simply weren't in her plan.

Donald would not listen. The thought of not having a family of his own saddened him so much that he couldn't believe that what Jean was saying was true.

A FEW MONTHS before the wedding, in May 1967, Donald was in the middle of one of his routine visits to the campus psychiatrist, talking about falcons. Staring at an abstract design on a card, he said that he saw a cliff with a hole in it. Through that hole, he said, there was a nest—a place where he could find newborn birds to take home and make his own.

A mysterious, dark, birth-canal-like passageway, through which Donald could find a new family: The Rorschach test had only just begun, and Donald was already giving the psychiatrist plenty to work with.

He looked at the second image and thought about temptation. He saw a woman ready to have sex with a man, and the man, according to the doctor's notes from the session, "suffering mental anguish as to whether he should or shouldn't." The man finally decided "to keep own values high" and not have sex.

The third picture reminded Donald of a friend of his, a beatnik. "He's on dope, I guess—he's unconscious."

The fourth and fifth made Donald think about a father and a son. He saw a son in bed, and his father coming to say good night. The father, he said, was going to walk out the door. Then he saw a son crying on his father's shoulder, asking his father for help. The son had done something wrong, Donald said, and the father was going to offer his son some guidance.

When he saw the sixth, all at once a violent drama unfolded in his mind—a man contemplating revenge, and a woman talking him out of it. "He's half listening and half not," Donald said.

The seventh, to him, was another revenge scene. This time, a son was avenging his father's death. The son, he said, "feels right in what he did, because the other person committed injustice to him and his family."

In the final picture, Donald saw himself.

"I'm climbing up a cliff," he said. "I'm at the top, and falcons are diving at me."

CHAPTER 8

While Donald was struggling at Colorado State, Jim, the maverick second son, spent a year after high school attending classes at a local junior college, rebuilding his academic record. To everyone's surprise, he did well enough to transfer the next year, 1965, to the University of Colorado at Boulder. It was lost on no one, least of all Jim, that his new college was better than Donald's. When it came to him and Donald, Jim never stopped keeping score.

Jim was about two years into Boulder—and a fixture at several bars in town—when he met Kathy. He was twenty, and she was nineteen. He spotted her at a dinner-and-dancing club called Giuseppe's. She was with an old high school friend, and Jim asked to cut in. Then he called her at her parents' place, where she was living, and they started dating. Early on, Kathy had picked up on the contempt that Jim felt for both of his parents. "They kept having babies and didn't deal with the younger ones," he once said. And he would rant about how much he detested his older brother—how Donald had been the big hero in high school, and Jim never really measured up. Now all that seemed to be behind him, she thought, or at least it ought to be.

When Kathy got pregnant, Jim didn't think twice before ask-

ing to marry her. For Don and Mimi, this outcome might not have
seemed ideal—even if, truth be told, they had done very much the
same thing when they were about Jim's age. It was pointless, in any
case, to say anything. This was Jim—he was going to do what he was
going to do. And after having just blessed the union of Donald and
Jean, they didn't have a leg to stand on.

The wedding took place a year after Donald and Jean's, in August 1968. They moved into a small red-brick bungalow downtown
and sometimes invited Jim's younger brothers over—but not Donald,
never Donald. While Kathy got along well with the other Galvin boys,
things were strained with Jim's mother, whose visits seemed more like
inspections. "You haven't dusted," Mimi would say, to which Kathy
would reply, "I don't have the time. Here's my dust rag if you want it."
Jim loved that.

Kathy gave birth to a son, Jimmy, who was just a few years younger
than Mimi's youngest, Mary. Jim dropped out of college and started
tending bar, a far cry from his goal of becoming a teacher like his
father. But that hardly seemed to matter: He was a family man now,
superior to Donald, he believed, in every conceivable way. When he
landed a regular gig tending bar at the Broadmoor Hotel, one of the
fanciest places in town, that seemed to throw off enough prestige to
make him feel as if he'd won.

JIM REVELED IN being a husband and a father, even as he took
every opportunity to break his marriage vows. A ladies' man before
his marriage, he had no interest in changing now.

One night, Kathy noticed his motorcycle outside a bar and she
went in, walked over to the table where he and his date were sitting,
poured a pitcher of beer on them both, and walked out. She wanted
to put Jim on notice—to let him know she had her pride.

Jim got back at her later, when they were alone. When Kathy decided to quit her day job and go back to school for a teaching degree,
he pulled the spark plugs out of her car to keep her from getting to
class. "Get a job," he said. When she got a ride from her mother, Jim
was waiting when she came home. He slapped her across the face.

As the violence increased, the worst thing she could possibly do,
she realized, was threaten to leave him. Once, when she tried that,
he punched her in the face so hard that she needed stitches. And she

never could bring herself to follow through on that threat. Every time Kathy was about to leave, she thought that maybe he'd get better, or that their son needed a father. On the few occasions when she did work up enough nerve to get out of the house, just for a night or two, Jimmy would say, "I want Daddy home."

There was another reason why Kathy wouldn't leave. She had started to notice that Jim seemed tormented by something that had nothing to do with her—something that made her almost feel sorry for him. He would hear voices. "They're talking to me again," Jim would say. His voice tight with emotion, he'd describe them—people spying on him, people following him, people at work conspiring against him.

Jim stopped sleeping. He spent his nights standing over the stove, lighting a burner and turning it down and then off and then lighting it again. In these states, he would act impulsively and violently, not toward Kathy or their son, but toward himself.

Once, walking in downtown Colorado Springs, Jim rammed his head into a brick wall.

Another time, he dove into a lake, fully clothed.

Jim's first hospital stay for a psychotic episode was on Halloween night in 1969, when little Jimmy was still a baby. He was admitted to St. Francis Hospital, but left within a day. Kathy was frightened for herself and her son. But she was also terrified for Jim. He was still her husband and her son's father, and leaving him now seemed impossible.

Kathy never liked Jim's parents—Jim himself had seemed to prefer it that way—but she felt Don and Mimi had to be told about what was happening. She could hardly believe their reaction. She had expected tears, maybe a show of compassion, or at least sympathy. Instead, Kathy saw two people trying hard to pretend the conversation wasn't happening at all—and, when pressed, questioning the premise of that conversation. Was everything really happening the way Kathy said it was? Jim's parents never came close to accepting that their son was entirely at fault, or even in danger. Instead, they framed what was happening as a marital problem between the younger couple—something that Jim and Kathy ought to try to resolve on their own.

The most remarkable thing, in hindsight at least, might have been what Don and Mimi did not say: that Jim's brother Donald had been exhibiting strange behavior, too. They weren't telling anyone about Donald, and they weren't going to start with her.

After talking with them, Kathy took Jim to visit with a priest—something Don and Mimi had recommended—but nothing came of that. One night, when Jim seemed completely helpless, Kathy finally took him to the University of Colorado Hospital in Denver. He stayed for two months, then came home. Jim agreed to get counseling on an outpatient basis at Pikes Peak Mental Health Center in Colorado Springs. A doctor prescribed medication for him, and he stabilized long enough for there to be some hope.

Only now and then would he lose his temper and hit Kathy again. Once, a police officer showed up, and Kathy declined to file charges. Another time, one of their neighbors called the police, and the cop escorted Jim out of the house. But he came back eventually. For better or for worse, he always did. And in the years to come, Don and Mimi never intervened. "Except the times Jim would leave, and he would go back and live with them," Kathy recalled, "which was fine with me. And then he would show up on my doorstep again."

ON A SPRING day in 1969, all twelve Galvin children gathered together in relative peace and harmony to honor their father at a com-

All twelve children and Mimi with Don, receiving his PhD, 1969

mencement ceremony at the University of Colorado. At the age of forty-four, Don had finally earned his PhD. The snapshot documenting this day is one of the only photographs in the Galvin family's collection in which all dozen children and both parents are pictured. Don is in a cap and gown, his hair already going gray. Mimi is by his side in a cream-colored spring dress with a canary-colored scarf, her hair back. The girls, Margaret and Mary, are in front of their parents in matching white dresses. And the ten boys are all together to their right, lined up in two rows, standing straight as bowling pins.

Jim is in the back row, fourth from the left, his dark hair tousled, face pale and sweaty. In the years to come, Mimi would point at this picture and say that this, one of the family's last uncomplicatedly happy days, was the moment when she first really absorbed the idea that Jim was in deep trouble—not just a maverick, the way he'd always been, but losing his mind. Like Donald.

On a spring day during the Great Depression, in a bustling town some-where in America, a squabbling, unhappy married couple welcomed into the world four identical girls—quadruplets. The press rushed to cover the story of the births, and the parents, whose resources were severely limited, allowed one of the local newspapers to hold a contest to name the four sisters. They also fielded offers of sponsorships from local dairies eager to use the girls to sell milk, and charged admission to visitors hoping to catch a glimpse of the babies at home.

Money did not solve the family's problems. One of the daughters had a psychotic break when she was twenty-two. The others followed, one after another. By the time they were twenty-three, all four sisters were diagnosed with schizophrenia. And in the early weeks of 1955, these four women—quadruplet sisters, twenty-five years old with identical DNA—were referred to the National Institute of Mental Health in Washington, D.C.

The psychiatrists at NIMH understood the rare opportunity that these sisters presented. By their calculations, quadruplets with schizophrenia were likely to occur only once in every 1.5 billion births. They entered the care of David Rosenthal, a psychologist and researcher at NIMH who, thanks in part to the quadruplets, would go on to become one of the century's most prominent schizophrenia researchers focused on the genetics of the illness.

The sisters stayed at NIMH for three years and Rosenthal and his team of two dozen researchers studied them for five more, protecting their privacy with pseudonyms. They gave them the last name Genain, from a Greek phrase meaning *dire birth,* and first names starting with letters corresponding to the acronym NIMH: Nora, Iris, Myra, and Hester. The city they lived in was never disclosed, and their parents became known as Henry and Gertrude. And in 1964, the year that the Galvins were settling into their new home on Hidden Valley Road, Rosenthal published *The Genain Quadruplets,* a six-hundred-page study of familial schizophrenia that would become a classic of the genre—a case study that, with its scrupulously nuanced take on the nature-nurture question, became every bit as consequential to the study of schizophrenia, it was said at the time, as the case of Daniel Paul Schreber.

BY THE TIME the Genain sisters came to NIMH, the search for a physical or genetic marker for schizophrenia had fallen out of vogue in psychoanalytic circles—out-argued, it seemed, by a new generation of therapists, Frieda Fromm-Reichmann among them. But in a separate silo—university laboratories and hospitals out of reach of the psychotherapists—neurologists and geneticists spent the 1950s and 1960s continuing the search for a biological marker for schizophrenia. The gold standard in such work was the study of twins. There could be no better way, it seemed, to test the hereditary strength of any condition than by seeing how many identical twins share the illness and then comparing that to the rates of disease in fraternal twins. Researchers in Europe and America conducted and published many major twin studies of this sort, starting with Emil Kraepelin in 1918 and continuing with others in 1928, 1946, and 1953. Each of these studies offered data showing a hereditary element existed, even if the numbers weren't overpowering. And each time, the response from psychoanalysts was more or less the same: How do you know the disease wasn't passed through families because the family environment was what caused the disease? How do you know it wasn't their mothers?

At NIMH, David Rosenthal believed right away that the very existence of quadruplets with a shared mental illness could settle this argument once and for all. "When one first learns that the quadruplets are both monozygotic and schizophrenic," wrote Rosenthal, "one can hardly help but wonder what further proof . . . anyone would want to have." But he also knew it was not that simple. In his writings about the case, he noted that many psychotherapists, including some of his own colleagues at NIMH, were unpersuaded. The parents of the Genain sisters presumably treated each girl pretty much the same: They dressed them alike, sent them to the same schools, set them up with the same friends. It would be every bit as likely, they argued, that these girls all had schizophrenia because the parents brought them all up the same way.

Rosenthal and his colleagues went to work collecting a family history of the Genains and found at least one instance of mental illness. The sisters' paternal grandmother had apparently had a nervous breakdown as a teenager, experiencing symptoms that one NIMH caseworker believed sounded like paranoid schizophrenia. But genetics only tells part of the story of any identical sibling, and the Genain sisters indeed were, in certain respects, different from one another. Nora was the firstborn and sort of the spokesperson for the group, the best piano player with the highest IQ, though she was given to tantrums. Iris, meanwhile, was described as "vacuous," but helpful around the house and a skilled beautician, while Hester was quiet, sober and retiring, "unkempt," as Rosenthal described her, "in a cinderella-by-the-fire fashion." Myra had a more "sparkling" personality, but paradoxically something about her affect seemed flat, as if she was playing the part of a person and not sure exactly how to do it. From an early age, the girls' mother had tried to separate Nora and Myra from Iris and Hester because she thought that Nora and Myra were brighter than the other two, whom she called "duller."

Then came the question of their home life. The more the researchers learned, the stranger it seemed—first peculiar, then appalling. Both parents were abusive. The father drank, had affairs, and was said to have molested two of the daughters. When the mother, for her part, discovered two of the girls engaged in mutual masturbation, she put them in restraints at night, gave them sedatives, and eventually forced them both to undergo female circumcision. In the view of the NIMH researchers, Gertrude was the same sort of mother that Frieda Fromm-Reichmann and Gregory Bateson had described—so controlling and anxious that her daughters had to have been traumatized by her in some way. "It is easy to see that the longer her family remained sickly and unwell, the more prolonged would be her gratifications," Rosenthal wrote. "Her house was her hospital."

In the end, nothing about the Genains' childhoods had been close to normal—not their schooling, and certainly not their sexual development. Even Rosenthal compared the experiences of these girls to the "extreme situation" concept developed by the Holocaust survivor and trauma theorist Bruno Bettelheim, in which one finds oneself overpowered by an inescapable situation, unprotected, never out of jeopardy. "Almost from the moment the quads were brought home from the hospital, an atmosphere of fear, suspicion and distrust of the outside world permeated the house," Rosenthal wrote. "The blinds were drawn, a fence erected, and the guns kept at the ready, with Mr. Genain patrolling. . . . The dread of kidnapping was constantly with them. . . . Threat was everywhere."

The nature of their childhoods seemed to corrupt the experiment. Certainly the researchers at NIMH would have had a more compelling nature-nurture experiment if the Genains had been a little more like, say, the Galvins—a more mainstream, middle-class family.

Even so, Rosenthal felt comfortable crediting a mixture of genetic and environmental factors for what had happened to the Genains. He rejected the argument that one single gene must have caused the illness, but he also rejected the belief that only the environment was to blame. In *The Genain Quadruplets,* Rosenthal became one of the first researchers to suggest that genetics and the environment might interact with each other to produce the symptoms of schizophrenia. And he started to outline what future work on this subject could do to break the impasse and broker a compromise.

"We must be more circumspect yet more precise in our theory-building," Rosenthal wrote. "Those who emphasize the genetic contribution seldom consider in earnest the role that environment might play, and environmentalists usually pay lip service to the idea that hereditary factors may eventually have to be considered as well." Future research, he declared, needed to build a bridge between the two ideas. "Both heredity and environment," he wrote, "are, of course, everybody's business."

Rosenthal's conclusions satisfied neither side. And yet he held on to this idea of nature and nurture commingling. He had no way of knowing how long it would take for that idea to catch on. But he came away from his time with the Genains determined to prove that the wellspring of madness might not be nature or nurture, but a fateful combination of the two.

CHAPTER 10

Donald went back and forth with the psychiatrist about whether his marriage to Jean was a happy or unhappy one.

One moment, he'd talk about the fun they'd had together on a six-week camping trip to Mexico. The next, he'd be admitting that things had not gone right ever since the wedding. In the three years that had passed since then—it was June 1970 now—Donald had come to believe he'd married Jean when he was on the rebound, after being rejected by his previous fiancée, Marilee, and that their life together now could barely be considered a marriage at all.

It was a sad story, but Donald wasn't telling it that way. Instead, he came off as stubborn, detached, critical, cool, and even slightly paranoid. The doctor, a psychiatrist from Colorado State University Hospital in Fort Collins named Tom Patterson, noticed a certain rehearsed quality to Donald, a rigorous self-control that flattened his entire personality, as if he were trying to keep the lid on something explosive inside of him. "He watches your every move," he wrote.

Both Donald and Jean were out of college now but still living in Fort Collins. Donald was working as a research assistant and taking classes in anatomy and physiology, still dreaming of a medical career one day, while Jean was completing her master's degree. That day,

Donald

Donald said that he came to the campus counseling center because someone he knew had recommended that he find a sensitivity group to help him communicate better with his wife. Before long, he revealed the real reason he was there: Jean told him that she was going to leave him in three weeks.

Donald spoke candidly with Patterson about how bad things had been recently. Jean had complained that he was distant most of the time, and that the rest of the time he was downright threatening. Where she once was the one refusing sex, now Donald only agreed to sex when she demanded it, about once a week. They ate separately and slept in separate bedrooms. He owned up to being withdrawn around her, and sometimes threatening her, too, but it was too late. Jean seemed fed up with him—and now that she had a paid assistantship lined up in a doctoral program at Oregon State University in the fall, she no longer needed him to support her. "In other words," Patterson wrote, "the marital relationship is a lousy one, with each person going their own way."

As calm as Donald seemed, Patterson knew all about the various other therapists Donald had seen since his first run through the bonfire years earlier. He even recalled once seeing Donald's Rorschach test, which he remembered as "quite pathological." On that day in June, the psychiatrist tried going a little deeper with Donald, moving beyond the pressing issue of his marriage to talk more about himself.

Their talk quickly became a full-fledged therapy session. Donald told the doctor that for years he had not been himself at all, but rather a mirror of what other people wanted him to be. He said he made a practice of reading people's facial expressions, gestures, and words for hints of the best way to react. He called his mad dash through the campus bonfire a plea for attention, and he said that he'd lied on a lot of the psychiatric examinations he'd taken. Recently, he'd said, he'd gone on an Eastern philosophy kick; he'd fasted for four days, and he bragged now about weighing just 158 pounds. The doctor was not impressed. Whatever Eastern terminology Donald was throwing around now in his conversation, the doctor believed simply made him seem more bland—not insincere, but not genuine, either. It often seemed to Patterson that Donald was about to cry, but then gathered himself and stopped before the tears came.

The psychiatrist came away believing that even if Donald had once displayed elements of paranoid schizophrenia, and even "may have done very bizarre or violent things," he might not be quite so far gone now. "He is in good contact with reality," he wrote in his notes. "He is evasive and probably does not commit himself to a deep relationship with anyone. . . . He has a low frustration tolerance and easily gives up on people or situations that threaten him." The doctor wondered if Donald's pent-up emotion was the result of him repressing his own desires and needs for too long—a theory that, strangely, almost seemed to blame Jean for Donald's problems. "He has given in to her needs, her wants," he wrote, "and has suppressed his own feeling so severely and for so long that he has difficulty now in expressing his affect."

Patterson ended the session by inviting Donald to come back the next day to talk some more. Donald did, and when he returned, he seemed strangely transformed—relaxed, even happy. He said that he and Jean had talked, and that she had removed the deadline when she learned that Donald was now seeing a therapist. They even went out to dinner together, and Jean agreed to try couples therapy.

Patterson was encouraged, but now that Donald wanted something from him, he felt ready to ask something in return. He said that he'd consider conducting couples therapy with Donald, provided Donald gave him permission to go through his file and learn more about his psychiatric history.

Donald darkened a little. He told the doctor he didn't believe in psychological tests. He thought the tests done on him were invalid, he said, and he wasn't sure anything in his file would be helpful.

"Therapy will be difficult because of this," Patterson wrote. "Can he be reached without denial?"

On his way out the door, Donald warily agreed to take home a paper-and-pencil personality test.

COMPLETE THESE SENTENCES TO EXPRESS YOUR REAL FEELINGS. TRY TO DO EVERY ONE. BE SURE TO MAKE A COMPLETE SENTENCE.
I LIKE: *Falconry, sex, swimming, travel, skiing. Communicating*
BACK HOME: *is a good place to visit for a short time.*
MEN: *should be more flexible in their thinking.*
A MOTHER: *should care for the development of her children.*
I FEEL: *tense.*
MY GREATEST FEAR: *not sticking to what I originally wanted.*
IN SCHOOL: *there's the best time of a life.*
I CAN'T: *Say "I quit."*
SPORTS: *develop character*
WHEN I WAS A CHILD: *I still am*
I SUFFER: *From self pity (not much)*
I FAILED: *chemistry*
SOMETIMES: *I don't care enough*
WHAT PAINS ME: *most are other people.*
I SECRETLY: *want to be happy when I'm alone.*
I WISH: *too much.*
MY GREATEST WORRY IS: *deciding what to do.*

NOT QUITE A week later, on a Friday night in June, Donald and Jean had another fight. It was all the same conflicts all over again, but worse, more fraught than before. Things were bad enough that Jean walked out of their apartment. Donald followed her, and found her nearby, sitting low to the ground, near an irrigation ditch. Either she was trying to have some time alone, or she was trying to hide from him. But once he found her, Donald started talking about how he wanted to drown her.

Jean talked him out of it. They both made it back into the apartment, more or less together, though Jean did make one thing clear: She would be moving to Oregon without him.

The next day was Saturday morning. Donald was still upset about the fight—and about Jean's decision to leave him after all. He took some mescaline, an experience that he later said not only offered him incredible insight, but helped him come up with the right response, the perfect plan.

That night—June 20, 1970—Donald came home with two cyanide tablets, procured, most likely, from a lab at the school. Donald dropped them into a glass of hydrochloric acid, took hold of Jean, and tried to hold her still—both of their faces above the glass as the cyanide misted into a gas.

The plan was for them to die together.

DONALD WAS A no-show for his next appointment. When Patterson opened the newspaper on Monday morning, he learned why.

> Fort Collins Police: 10:20am Donald Kenyon Galvin, 24, of 27G Aggie Village was booked for protective custody in connection with an alleged suicide and possible homicide attempt. He was being held in city jail this morning on authority of the district attorney. He was first taken to the Colorado State University Student Health Center for treatment.

Donald's plan hadn't worked. Maybe he loosened his hold on Jean, or maybe his grip was never that strong to begin with. But she tore herself away, ran from the room hysterically, and called the police. After reading the report in the paper, Patterson found Donald at a hospital, where he'd been sent on a "confine and treat" order while the district attorney's office decided whether to charge him or have him committed. Much to the doctor's alarm, Donald still hadn't seemed to have come down from the experience. As Donald talked, he came off as euphoric, even boastful—an unmasked comic book villain, crowing about how he'd fooled everyone for years. He talked about the time that he killed a cat, but this time instead of being terrified, he was almost gloating. He said he'd recently dismembered a dog in the bathtub, too, just to upset Jean.

Nothing in Patterson's notes from Donald's sessions suggested he was capable of anything like this. Had Donald deliberately pulled the wool over Patterson's eyes, or had he simply fallen apart without any real warning signs? Had the doctor missed something violent in him? Had he been too willing to have faith in him?

That, at least, was over. Donald had a new diagnosis. "He is probably an intelligent paranoid schizophrenic," Patterson wrote, "who has wide mood swings from elation to depression. . . . I think the in-patient commitment procedure is definitely the right thing to do."

—⚏—

The Colorado State Hospital in Pueblo is a collection of large, bland brick buildings at the center of a town that has sprung up around it, largely to accommodate the growing staff of health care workers serving the expanding patient rolls. When the hospital first opened with about a dozen patients in October 1879 under a different name, the Colorado State Insane Asylum, the facility was just a farm-house, and Pueblo was a sleepy town on a flat stretch of desert, a hundred miles south of Denver. The institution got its new name in 1917, having grown by then to treat more than two thousand patients—each one housed there with very little hope that they might ever be released.

The early patients at Pueblo were subject to a seemingly endless array of chemical and electric treatments designed to pacify them. In the 1920s, as the eugenics movement gained momentum, Pueblo's doctors sterilized their female patients, despite lacking the legal authority to do so. It never seemed to occur to any of them that it might be a bad idea. "We considered it a minor operation," the hospital's longtime superintendent, Dr. Frank Zimmerman, said years later. "So they will not produce more mental deficients."

By the 1950s, the hospital housed more than five thousand patients, becoming a small, largely self-sustaining community—bigger than the county seat of the biggest county in the state—with parents and children and grandchildren all going to work there at the same time. Unable to rely on the state legislature for funding, the hospital arranged for patients to grow their own crops and operate a dairy farm, a pig farm, a garden, and a factory where the patients made textiles. Pueblo had become a colony for the mentally ill, where people stayed forever; the most popular treatments in those days were electroshock therapy for depression, insulin coma therapy for schizophrenia, hydrotherapy for mania, and fever therapy for tertiary syphilis.

Only after institutions like Chestnut Lodge changed the thinking about mental illness did the brutality at Pueblo and other state hospitals start to become a subject of debate in the broader culture. One of the earliest and most powerful exposés was *The Snake Pit,* a 1946 semi-autobiographical novel by Mary Jane Ward—later made into a movie starring Olivia de Havilland—about experiencing scalding hot baths and electroshock therapy as a patient in a state psychiatric hospital in New York. In 1959, the Colorado State Hospital in Pueblo also became the subject of a book, a provocative roman à clef called *The Caretakers,* written by a former employee named Dariel Telfer. If you set aside its more sudsy, *Peyton Place*–like aspects, *The Caretakers* presented a vivid picture of some of the more popular treatment practices of the time: shock therapy, Thorazine, tranquilizers, solitary confinement, sodium luminal, sodium amytal. One character's cavalier description of a high-security ward at the hospital is especially telling: "These are mostly psychopaths. They can do anything they've a mind to. Mostly they want sex and good times and liquor. They need to be kept busy on account of when they got nothing to do, they get meaner'n hell. They oughta be put to work, every single one of 'em. I got one on my ward that's been down in restraint two weeks. According to her chart, she's had over two hundred shock treatments. Over two hundred! Thinka that!"

A *New York Times* reviewer called *The Caretakers* a clarion call for investigation and reform. Sure enough, in 1962, a Colorado grand jury delivered a scathing thirty-page attack on the hospital in Pueblo, revealing many of the same problems that had been depicted in *The Caretakers:* neglect and abuse of patients; unlicensed doctors (at least one of them drunk on the job); patients escaping and running wild on the grounds. The occupational therapy school had become "the center of immoral activity"; one shady section of the hospital grounds where patients would meet to have sex had become known as "Bushville." In one case, an illness reported on a Monday was not acted on until Saturday; that patient subsequently died.

Reform, it turned out, was just around the corner. President John F.

Kennedy's Community Mental Health Act of 1963—inspired, in large part, by the Kennedy family's tragic experience lobotomizing and institutionalizing the president's eldest sister, Rosemary—ordered the downsizing of large institutions like Pueblo. This was supposed to be good news both for the people who had been unnecessarily warehoused and the harder cases who could use more individualized attention. It didn't exactly work out that way. At the same time that the federal government was emptying out large institutions for the mentally ill, the doctors at Pueblo had gone all-in on the new, miraculous neuroleptic drugs that could treat the mentally ill without expensive person-to-person contact.

These drugs, the most consequential advancement in the treatment of psychotics in the twentieth century, had arrived a decade earlier, well outside the field of psychiatry. In 1950, a French surgeon named Henri Laborit was working on a new type of battlefield anesthesia that mingled narcotics with sedatives and hypnotic drugs. The drug, which he called chlorpromazine, had its first human trial in 1952. As Laborit described it, patients on his new drug developed a "euphoric quietude," becoming "calm and somnolent, with a relaxed and detached expression." Laborit himself even likened the effects of the drug to a "chemical lobotomy." Chlorpromazine debuted in the United States in 1954 under the brand name Thorazine.

In the years that the Galvin boys were coming of age, Thorazine was becoming widely accepted as a sort of miracle drug, able to calm patients out of psychosis when nothing else but surgery or shock therapy would have done the trick. By the time Donald was committed and sent to Pueblo, in 1970, more than twenty drugs had entered the market, all variations of Thorazine. For large state-run hospitals like Pueblo, medication promised to deliver what therapy had seemed unable to—fulfill the Kennedy-era vision of mental health treatment, stop the warehousing of these patients, and help some or even many of them leave the hospital. But Thorazine was no cure—it reduced some symptoms, but at best forced an unsteady truce with the illness itself. And from the start, there were questions, starting with side effects: tremors, restlessness, loss of muscle tone, postural disorders. What Laborit saw as calm and somnolent seemed to others more like muzzled and muffled—a knockout punch. Some patients never seemed to come out of their pharmaceutical stupors, and if they went off the drug at any point, the next round of psychosis tended to be more acute than the last. And perhaps the biggest question of all: How did it work?

Even today, no one knows for sure why Thorazine and other neuroleptic drugs do what they do. For decades, doctors have been treating schizophrenia pharmacologically without a clear understanding of the biology of the illness. At first, the best that researchers could do was examine what Thorazine does to a patient's brain and extrapolate theories of the illness based on what they noticed. The first credible theory came in 1957, when a Swedish neuropharmacologist named Arvid Carlsson suggested that Thorazine treated the symptoms of schizophrenia by blocking the brain's dopamine receptors, stopping many of those hallucinogenic, deranged messages from spiraling out of control. Carlsson's work formed the basis of what, among schizophrenia researchers, became known as the "dopamine hypothesis"— the notion that overactive receptors somehow caused the disease.* The problem with the dopamine hypothesis was that another neuroleptic drug, clozapine, emerged that alleviated some of schizophrenia's symptoms even better than Thorazine, only it worked on those same dopamine receptors in seemingly the exact opposite way—increasing dopamine levels where Thorazine had inhibited them. If two effective antipsychotic drugs were sending dopamine levels in different directions, something besides the dopamine hypothesis had to be explaining why they worked.

Practically every drug prescribed for psychosis, from Donald's time until now, has been a variation on Thorazine or clozapine. Thorazine and its successors became known as "typical" neuroleptic drugs, while clozapine and its heirs were "atypical," the Pepsi to Thorazine's Coke. Like Thorazine, clozapine could be dangerous: Concerns over drastically low blood pressure and seizures were serious enough to take it off the market for more than a decade. Even so, drugs became the common treatment of schizophrenia, and the psychiatric profession's great schism only widened. On one side of the street, doctors at

---

* Years later, Carlsson would collaborate on the first selective serotonin reuptake inhibitor, or SSRI, to reach market, a precursor to Prozac. The impact of his dopamine work on treatments for Parkinson's disease earned him the Nobel Prize in 2000.

the large state hospitals said schizophrenia required drugs, while the therapists in more rarefied settings still recommended psychotherapy.

Like most families, the Galvins were at the mercy of what was a mental health care system in name only, forced to choose from options they weren't equipped to assess. In the end, their decision came down to money. While insurance paid for the dependents of Air Force personnel, Donald was twenty-four now, and no longer covered. And so the decision was made for them. Pueblo was his only option.

DONALD CAME TO Pueblo after six days in jail waiting for the commitment to come through. That gave him six days to become increasingly terrified by the prospect of being committed to a mental hospital. In his intake interview, he tried to say that he had a perfectly reasonable explanation for what he'd just tried to do to Jean and himself with the cyanide: He'd taken peyote for the first time a few weeks earlier, he said, and later heard that the peyote could have been LSD. He said that he was fine now, and that he would let his wife leave him without any objections; he'd been "uptight" like this once before, he said, when his first fiancée had left him, and he'd gotten over it then, too.

The doctors at Pueblo were wary. "Psychotic episode should be considered," the notes read. "Diagnosis: depressive neurosis—or psychotic depressive."

The next day, during another visit with doctors, Donald gripped the table as he insisted he was well, and ready to stand on his own two feet. He did not want to be institutionalized—that much was clear. While the doctors did not necessarily believe him, they also weren't sure, the cyanide incident notwithstanding, exactly how sick he was. Donald received a new diagnosis: "anxiety neurosis, moderate to severe, with obsessive features."

By Donald's arrival, Pueblo had retrenched from its peak of six thousand patients to something more like two thousand. And yet with still only a handful of real doctors tending to the patients, the standard of care hardly improved. The staff members caring for the patients mainly were called "psych techs," people with basic nursing training but often no nursing degree. Their main responsibility was dispensing Thorazine, Haldol, and other meds—the substitute for a doctor's care. The pills were brought to the wards in bulk, and the psych techs would pass them out to patients, often at their discretion. "It was like passing out snacks," remembers Albert Singleton, who spent decades as the hospital's medical director.

Donald was prescribed Tofranil, an early-generation antidepressant with harsher side effects than the selective serotonin reuptake inhibitors, or SSRIs, of the Prozac era, and Mellaril, a first-generation antipsychotic drug in the Thorazine mold that was eventually pulled from shelves when it was found sometimes to cause cardiac arrhythmias. And a few weeks later, on July 15, 1970, after cooperating with his treatment, Donald was released from Pueblo. Thanks to his psychiatric commitment, he was facing no further jail time.

While he was in the hospital, Jean had filed for divorce.

WITH DONALD BACK home on Hidden Valley Road, Don and Mimi faced a choice: Should they stop everything and stay home with their sick son? Or should they give him a chance to fend for himself, and continue traveling as a couple to events with the Federation?

In the end, they didn't feel like they had a choice at all. The Federation was not just their only chance at the life they always wanted. It was the family's only source of income. If Don and Mimi didn't keep up appearances—if Don came alone to Santa Fe or Salt Lake City, and made it known that they were struggling with an adult son's illness, and that son was at home and that his marriage had failed—it would have raised so many other questions that they were not willing to answer that they never seriously considered changing a thing.

Instead they helped Donald find a job in the admissions department of a business school in Denver. Donald was sent to North Dakota to recruit students, leaving town long enough for Mimi and Don to fly to Salt Lake City in September for a gala featuring Ballet West, and then again in November for a luncheon honoring the ambassador from Argentina, Pedro Eduardo Real, and his wife. "I sat next to the consular officer from Mexico City," Mimi wrote her mother, on the hotel stationery. "He and Donald and his wife spoke in Spanish, and enjoyed one another very much." Mimi went on to boast about Don dispensing $75,000 in grants for the symphony, the ballet, and other groups. "You should be very proud of his good works in so many fields!" She closed the letter by talking about the girls: "Mary C. and Margaret especially want to see you. They are growing up so quickly

and this year may be the last that we would have everyone here to see
you at one time!!!"

Donald's hospitalization—his attack on his wife, the divorce,
Pueblo, the prescriptions—went unmentioned. Mimi dared not say
a word.

DONALD'S TRIP TO North Dakota brought him nowhere close to
Oregon, where Jean was now living. But that did not stop him from
turning the trip into an excuse to travel more than a thousand miles
farther west to try to speak face-to-face with the woman who was
divorcing him. He and Jean spoke for five minutes, long enough for
her to tell him that she wouldn't see him. His uncle Clarke, who lived
not far away, got him and brought him home.

Back on Hidden Valley Road, Donald took to declaring that
his marriage to Jean was still in existence spiritually—because, he
explained, the Church had never signed off on the divorce. He announced that he wanted to become a priest, and applied to the chancellery, which sent some people to visit him. After a few minutes of
watching Donald talk a mile a minute about his dream of construct-
ing a new church to honor St. Jude, the meeting was more or less over.
Donald never heard back from them.

One afternoon, Margaret, eight years old, came home from
school to find Donald naked and shrieking. She looked around and
saw that the house was completely empty. Her brother had carried
every single piece of furniture out of the house and stashed them in
the hills. Margaret remembered the look of distress on her mother's
face as she told her to go lock herself inside the master bedroom—
the only room in the house with a lock. She remembered finding
five-year-old Mary, already there, waiting for someone to keep her
company. A few moments later, their mother joined them. Mimi said
they had to stay put while they waited until the police came to take
Donald away.

Through the closed and locked door, Margaret heard Donald
shouting biblical sayings, mixed with words with no meaning at all.
She remembered it taking forever for the police to come. Finally, she
heard the crunch of gravel on the driveway, and saw the red and blue
lights flashing against the bedroom walls.

She remembered her mother leaving the room to talk to the police, saying, "He is a danger to himself and others."

She remembered leaving the master bedroom and seeing her brother seated in the back of the police car—and the blue and red lights fading into the distance.

And she remembered him, sooner or later, coming home again.

CHAPTER 11

One bright Monday in June 1971, a jet plane landed at Sardy airfield in Aspen, Colorado, carrying seventy members of the Ballet West dance company. Each summer, the Salt Lake City troupe came to Aspen for a residency, performing to a friendly audience of well-off owners of beautiful second homes. This summer was different: Ballet West would be rehearsing and performing six new productions in advance of a late-summer European tour featuring a few guest stars: Linda Meyer of the San Francisco Ballet; Karel Shimoff of the London Festival Ballet; and, from the New York City Ballet, one of the finest male dancers of his generation, Jacques d'Amboise.

The airplane door opened. Out came the three guest dancers, glamorous and smiling. And up the metal staircase climbed a little girl wearing white knee socks and clogs and a gossamer dress, handmade by her mother. Margaret Galvin—just nine years old, with long dark hair parted down the middle and an impish smile—was carrying a bouquet of flowers for Jacques d'Amboise. She was part of the welcoming committee, happy to be chosen to hand over the bouquet on behalf of the group that had sustained Ballet West for years—an organization run by her father.

Don and Mimi's trips to Aspen with the Federation of Rocky

Mountain States were heaven for Margaret. She dreamed of nothing but dancing, of joining Ballet West when she was older; she even wore the same plain blue clogs preferred by the members of the company. She took classes in Aspen during the summer months—three a day, plus pantomime and tap—wearing an outfit her mother bought for her at an Aspen boutique. By the age of twelve, Margaret was being fast-tracked as a dancer, practicing from 7 a.m. to 3 p.m. every day in Aspen, and then going straight to rehearsals, then home for a quick bite before attending performances at night. When Margaret's sister, Mary, was old enough, she joined her on adventures in Aspen, taking walks up and down Maroon Creek, looking for mushrooms, and riding the chairlift together to Aspen Highlands. They both noticed how people sought out their father for conversation and counsel, and how relaxed and comfortable he was with everyone, rarely without a martini in his hand. Their mother seemed to enjoy it, too, even if, on many evenings, as Mimi dabbed on her Estée Lauder perfume, she'd fret to the girls that the family didn't have the money for her to have what she needed to wear.

And what about the boys? In the years before Donald went to Pueblo, he had been out of the picture, married in Fort Collins, at least two hours by car from Hidden Valley Road. Once he got sick, he was sometimes home, sometimes at the hospital, and sometimes attempting to live independently, finding jobs in stores or selling items door-to-door. As long as Donald was well enough to try living somewhere else, these trips to Aspen and Santa Fe could continue.

Jim was married, living with Kathy and Jimmy in downtown Colorado Springs. The next boys in line, John and Brian, were in college—and the next after them, Michael and Richard, were high school age and only came to Aspen and Santa Fe sometimes. The rest of the time, they stayed home and looked after the four youngest boys—Joe, Mark, Matt, and Peter—taking them to team practices, making sure they ate meals. They could take or leave these Federation excursions; they'd rather be on the ice or the ball field.

But for the girls, these trips away from home were everything. Margaret could pretend that she belonged there all the time. The spell would break whenever the brothers would come along with them. *You need to be away from here,* Margaret would think, watching Joe or Mark or Matt or Peter snapping towels or doing cannonballs in the pool. *This is my place.* The last place she wanted to be was with any of her brothers—not on Hidden Valley Road or anywhere else.

MARGARET HAD BARELY been a toddler in 1963 when the family first moved to Hidden Valley Road, and in those earlier, happier years, she existed mainly as a prop for her brothers. Each boy before her had gone through a version of this, too. "We were the football," her brother Richard once said, remembering being tossed around their old living room when he was the littlest. In the girls' case, first Margaret and then Mary became everyone's toy.

In close quarters, all ten boys tickled and teased her and hurled her through their spanking machine, for no reason other than it seemed to pass the time. This had thrilled Margaret at first. She had worshipped her brothers; she was two years younger than the youngest boy, Peter, and seventeen years younger than Donald, the oldest. Once she was big enough, Margaret would scramble through the scrub oak in their backyard and climb the pines to spy on the boys as they built a three-story tree fort at the top of the hill, overlooking the entire valley. When the boys finished the fort, Margaret was afraid to climb it, but when her brothers called her a sissy, she did it, anyway.

Margaret was too sensitive not to internalize the conflict between the brothers—all that wrestling and punching and brawling—even when it wasn't about her. And soon enough, it became about her. As she got older, Margaret became less of a mascot and more of a target, a sitting duck. On her way home from school, her brothers threw pinecones or water balloons at her from the top of the hill. Once she was home, the spanking machine remained fully operational—only now there would be obvious sexual undertones. Mark once was told by his older brothers that he had to run over and "do" Margaret. She would be groped and handled strangely, bullied harshly in a way that some of the boys might have considered innocent and fun.

Was this abuse? Or was it a bunch of wound-up athletic boys with no sense of limits, no internal regulators, getting physical with one another and her? Margaret would spend years wondering about that. In any case, she was too powerless to engage in open combat with them. She wanted to be comforted and protected. On Hidden Valley Road, home of the twenty-four-hour wrestling tournament, that never seemed to be an option.

Clockwise from top: Peter, Mark, Joe, and Matt

A generous portion of Margaret's, and later Mary's, formative years took place in the spectator section of the Broadmoor World Ice Arena, watching practices and games. The youngest four brothers formed their own little unit within the larger family, playing every sport together, with hockey their finest. Joe was mild-mannered and introspective. Mark was a chess prodigy, sensitive and, by Galvin standards at least, preternaturally well-behaved. Matt was prone to mischief, but also had a flair for making pottery. Peter, the youngest, was the family's great insurgent—more rebellious than any of the others ever had been, unable to tell Mimi and Don anything but "no." But barely a week went by without one of the four hockey brothers making it into the Colorado Springs *Gazette* for their performance in hockey games—culminating in one glorious moment, when three of them were all in high school together, all on the same team, and all on the ice together, and Joe and Mark both assisted on a goal scored by Matt, and the announcer cried, "Galvin to Galvin to Galvin!"

At home, the boys fired off sports trivia at one another between practices, and watched whatever game was on, and wrestled and fought. Even when Matt shattered his jaw and occipital lobe during one hockey game, and had to be rushed to the emergency room, and spent weeks with a constellation of pins and stitches keeping his head together, that, too, was typical Galvin fare, nothing out of the ordinary. Margaret sought shelter with her mother in the kitchen, helping her out as she listened to Mimi go on about the annoyances of the day. She would go to the market with her mother, controlling the second shopping cart that was necessary for holding enough groceries for a family of their size. And she would submit, obediently, to her mother's constant corrections of her behavior, her school performance, and her attempts at painting and drawing.

In sixth grade, a teacher complimented Margaret's artwork, and something registered inside her. Only when she was dancing had she felt anything like this—the sense that she might be able to create something out of nothing, to matter, to be more than just a piece of furniture in her brothers' playhouse. She had watched her mother with her watercolors, painting mushrooms and birds. Now she wondered if that was something she could do one day, too.

But Margaret was a little too cowed by Mimi to compete with her that way. She always wanted more reassurances and support and approval than her mother was willing to give. So she put those feelings on a shelf, for the time being.

DONALD HAD BEEN off at college when they'd moved to Hidden Valley Road, and had only come home for visits. After his release from Pueblo, his stay at home seemed open-ended—until he got better, maybe, or at least could be trusted to hold a job and live alone. That day seemed far off to everyone, and for Margaret, who was eight when Donald moved home, each day with him there brought the fear of something new. Donald would lead masses for a parish of one— himself—shouting the Beatitudes, the Hail Mary, and biblical passages. Later, he would go to the art store and buy some cheap picture frames and mount them to the wall, framing one-word quotes like *sincerity* all around the house. Too contained by the house, he would walk hundreds of miles around the neighborhood, county, and state.

At mass every Sunday, Mimi told the children to pray for Donald. But in public, she would titter and smile and say that their family of twelve children was a little daffy or eccentric or adorable—like the family in *You Can't Take It with You*. The most she would say about Donald was that he had not been the same since his wife left him. That woman had not been a good choice for Donald. The marriage was all wrong to begin with. Now he couldn't seem to get over her. "She was *not* a wife—she *wasn't*," Mimi would say, shaking her head—implying, without exactly saying, that her son's problems were the result of a broken heart.

As Mimi doubled down on her perfectionism, the girls became her most trusted deputies. Both girls tried to help their mother—taking out the trash, mopping the floor, washing the dishes, setting the table, vacuuming, cleaning the bathrooms—as if there wasn't a sick twenty-five-year-old man stalking the yard or writhing on the floor. Six o'clock remained the dinner hour, and whoever was home was expected to sit down and eat—even if, in the case of Donald, he had spent much of the day dressed in a monk's robe. Mimi also tried to include Donald in family outings, but the results were mixed. When she brought him to a hockey game, he got down on his knees in the middle of the crowd and started praying. That evening, as he chewed on a mouthful of steak, he announced to everyone at the table that he was eating his father's heart.

Hoping things might turn around for Donald did not seem to work in the slightest. Margaret turned nine, ten, and eleven on Hidden Valley Road with Donald dominating everything about their home life. Margaret and Mary got used to him exchanging blows with the brothers still at home—Joe, Mark, Matt, and Peter. Once, Donald thought one brother had made off with his medicine and tried to choke him. Another time, Donald took an entire bottle of pills and an ambulance came for him, again. The only person willing to break the silence around the problem of Donald was Jim, the maverick second son, who took pleasure in dropping by and saying what he was sure everyone else was thinking. *Shut the fuck up. Get out. Why don't you leave? Why don't you get out of here? What are you doing living here at your age?*

Jim came up with a nickname for Donald: Gookoid. That name stuck. Most of the younger siblings invoked the name more than once a day. Teasing Donald felt better than avoiding him, which drained them of all agency. Making Donald the brunt of their jokes gave them a sense of power over a situation they had no explanation for—and reassured them that whatever Donald was, he was not them.

ONE AFTERNOON, DONALD pulled a knife on Mimi. Margaret dashed to the phone in the kitchen and tried to call the police again—but this time, Donald lurched around and yanked the phone out of the wall. Margaret started wailing, sobbing. The wire from the phone had given her an electric shock.

Margaret watched her mother take control—ordering her daughter, one more time, to go into the master bedroom and lock the door behind her. Margaret did what she was told, but put her ear to the door. After what seemed like forever, she heard a scuffle in the kitchen, some shouting—voices of other people.

Joe and Mark had come home from hockey practice. They were confronting Donald, protecting Mimi—possibly, Margaret thought at the time, saving her life.

Donald stomped out of the house, vowing he would never go back to the hospital. Margaret heard nothing after that, except for the sound of her mother crying.

DON
MIMI

DONALD
JIM
JOHN
BRIAN
MICHAEL
RICHARD
JOE
MARK
MATT
PETER
MARGARET
MARY

CHAPTER 12

It was with no small measure of satisfaction—a declaration of victory may have been more like it—that Jim stepped in to help protect the youngest Galvins from Donald. Jim had often had all the younger boys and girls over to his house for sleepovers. He took Mary and Margaret to the movies and ice-skating and swimming, and skiing on the Broadmoor slopes, and riding on the Manitou Incline, a well-known funicular tourist attraction, where he had a job. He taught Margaret how to fly a kite and ride a bike. All the kids got rides on Jim's Yamaha 550 motorcycle.

When things were too strained at home, Mimi and Don were all right with the girls spending entire weekends at Jim and Kathy's house. Jim seemed on an even keel to them now, his stay at the hospital behind him. Kathy became almost like a mother to both girls, brushing and curling their hair while they all watched *Sonny & Cher.*

For the girls, it was an easy choice. They would much rather stay with Jim and Kathy if it meant avoiding Donald. To their parents, Jim was coming to the rescue, taking some of the burden away from them when they needed help the most.

Jim was so kind to the girls, so welcoming and accepting, that when he started to touch them, it almost seemed normal.

HIS APPROACHES WERE always the same. It would always be very late at night. Usually, he was drunk, after a shift at the bar. The TV would be on, and Kathy would be in bed, and he would come into the living room and lie beside Margaret on the green-flowered couch where she was sleeping. Margaret remembered the sound of bubbles from the fish tank, and the greenish blue damask pattern of the couch (a hand-me-down from Mimi), and the wicker rocking chair that was turned toward the kitchen, and the record albums standing in a row on the floor between cinder blocks, and the window looking out into the courtyard and toward another duplex, and the sound of the national anthem that played when the television stations went off the air. He'd penetrate Margaret with his fingers, and he'd try with his penis but could never accomplish it.

He had first gone after Margaret, as she remembered it, when she was about five—around 1967, a few years before Donald's first commitment to Pueblo, when she first started having the occasional sleepover at his place. She was too young to understand what was happening as an act of violence. Manipulation and attention and predation all mingled together until, with nothing else to compare it to, what was happening seemed a little like love. And so when the occasional sleepover turned into long weekends, this seemed natural to Margaret. Once, she was with Jim at a store that sold polished decorative stones, and she spent a lot of time looking at one called the tiger's eye. Jim bought it for her. For years she adored that stone—until the day, years later, that she finally realized just how wrong it all was.

Margaret's feelings about Jim started to change when she was about twelve, before she had her period. This was when she began fending him off at night, refusing him. Even then, she told no one about what Jim had been doing—especially not her little sister, Mary, who in Margaret's view seemed far too young to be allowed to know. What Margaret hadn't considered was that Jim would turn to Mary as soon as Margaret thwarted him.

Mary had been about seven, maybe eight, when she had a moment alone with her big sister and asked if she, too, had ever been bothered by Jim. Margaret's answer was short, definitive—a conversation-stopper. "I don't know what you're talking about."

It would be years before the sisters would talk about Jim again.

THE GIRLS WERE among the first to see how Jim was every bit as unstable as his brother Donald. Even beyond what he did to them at night, he was drinking too much all the time, and fighting with Kathy more and more. While Jim never hit them, they did see him hit Kathy sometimes, lightning-fast rampages that were so self-contained, it seemed almost as if he became someone else briefly, and then reverted back to Jim after that. Then Jim started having difficulty reverting. Mary remembered having to leave the house more than once with Kathy and Jimmy to get away from him.

In the calculus of their preteen minds, blocking out the nighttime encounters with Jim and his violence toward his wife was the price Margaret and Mary had to pay to gain a few days of liberty from the house on Hidden Valley Road.

It was more than that. Being with Kathy and Jimmy gave them a sense of belonging they couldn't get at home, not when so much attention was being paid elsewhere. They both so dreaded Donald that in the contest between Donald and Jim, Jim won. That, if nothing else, explained why they both kept coming back.

But there was another reason, too.

It is also true that they were too young to know for sure that what he was doing was not right—because Jim was not the first brother to attempt it with either of them.

One of Mary's first childhood memories, from about the age of three, was Brian molesting her. Margaret also remembered being touched inappropriately by Brian, more than once. Brian had been so well liked by them all, and he had left the house so quickly after high school, the girls never told anyone about Brian, either.

The truth about the Galvins—what Mimi and Don never saw, and never could have allowed themselves to see—was that by the time Jim advanced on the girls, everyone in the house on Hidden Valley Road seemed to be operating in a world with no consequences.

DON
MIMI

DONALD
JIM
JOHN
BRIAN
MICHAEL
RICHARD
JOE
MARK
MATT
PETER
MARGARET
MARY

CHAPTER 13

If Donald had been the imposing leader of the Galvin boys and Jim the resentful second-born, the third son, John Galvin, did his best to stay out of the fray entirely. The family's most devoted classical musician, he practiced intently, toed the line in school, and spent most of his time at home avoiding his older brothers. Once he left home in the fall of 1968 on a scholarship to the music program at the University of Colorado in Boulder, John had rarely come back to Hidden Valley Road.

In his junior year, in the fall of 1970, John fell in love, and with some trepidation he brought his new girlfriend, Nancy, also a music student, home to meet his family. From the moment they walked through the door, John felt like the visit had been a terrible idea. Everything was so much worse than it had been when he left. The whole household had turned in on itself. Where everyone once was out in the fields, flying falcons and climbing rocks, now they were hiding Donald from view as best as they could. He saw how his mother had an inventory of stock speeches, designed to counterprogram Donald's: a lot of talk about being Catholic, and more of her name-dropping and cultural one-upmanship, the old stories of Grandfather Kenyon, the new ones about Georgia O'Keeffe. With Donald talking to the devil in the garbage can or pacing and fidgeting and prattling on, they saw Mimi at her worst, trying to control the eight children who remained at home while denying, at least outwardly, that anything was wrong at all.

John and Nancy tried to keep things light. They played for Mimi, which delighted her—majorcas, Chopin études, and Beethoven sonatas late into the night. But in the way that new spouses sometimes give their partners permission to feel things they're ashamed to feel, Nancy was more vocal about what they were witnessing. She was from a small family—"normal-sized," she'd say—and could not stop remarking on how the house on Hidden Valley Road seemed like such an emotional shambles, drenched in confusion and anarchy. The endless fighting, the absence of personal space, four sets of bunk beds, no room for anyone to be alone: How could a mother be expected to raise that many children in such a pressure cooker? And those two little girls—how on earth did they have any privacy? How could anyone who lived there have a moment just to think?

When John looked at his parents, he saw two people trying hard to claw back some small part of what they'd once had. Their early years had been filled with such promise, and now so much was going wrong. This, John thought, helped explain why his father took him aside during one of his visits and suggested that he try to be more of a success than he already was—to give up on music and study politics. "Music is a selfish profession," Don said. "You spend a lot of time in a practice room. You don't socialize much. What good are you doing?"

His father's words saddened John, but he wasn't surprised. He had always been convinced that Don never thought much of him. He'd spent so much of his childhood in the background, he never thought anything he'd ever do would catch his father's notice, much less impress him. John was not alone in believing this about himself. Don Galvin was such a titanic figure in the lives of his sons—the falconer, the intellectual, the war hero, the classified intelligence officer, and now the counselor to governors and oil barons. All ten boys, in one way or another, grew up believing they could never be the man he was.

So no one was more shocked than John when, on the day of his wedding to Nancy in 1971, Don confided to the bride's mother, "She got the best of the litter."

Brian Galvin—the fourth son, after Donald, Jim, and John—
was the best-looking Galvin boy, even more handsome than square-
jawed, all-American Donald. Their father had nicknamed him the
Black Knight, for his jet black hair. He ran faster, threw a ball harder,
and his natural musical ability was leaps and bounds beyond the others', even his studious brother John. Once Don and Mimi saw that
Brian could listen to a piece of music on the radio and play it perfectly on the piano moments later—classical, jazz, blues, rock 'n' roll,
anything—they invested in private piano lessons for him.

For all his talent, Brian was also quiet, almost shy. He spent a lot
of time playing chess with Mark, the eighth son, who was six years
younger than he was—and happened to be a chess prodigy. But in the
way that children who withhold have that effortless way of attracting
the most attention from their parents, Brian's remoteness, his mys-
tique, made his parents want to please him even more. They were in
thrall to Brian's talent, too, and alarmed enough by young Donald's
emotional ups and downs to welcome any chance for the other boys
to be successful. And so when Brian and some high school friends

Brian, far left, with
his band

were forming a rock band, Don bought Brian a brand-new Höfner bass, just like Paul McCartney's.

The boys named their band Paxton's Backstreet Carnival, after a track from a Strawberry Alarm Clock record. They played covers: the Beatles, the Doors, Steppenwolf, the Stones, Creedence, the Zombies. Brian played bass and flute, and he was also the de facto bandleader, the one who could figure out the intricacies of any song in no time at all, score it in his mind, and then teach the song to the rest of the band. Over the summer break, he taught himself electric guitar, and by fall he'd taken that over, too. "In some ways, he was, I think, the most, more gifted of all of us," said Bob Moorman, the organ player and lead singer, whose father was General Thomas Moorman, the superintendent of the Air Force Academy.

Brian's band booked gigs all over the state: Glenwood Springs, Denver, South Trinidad. They played proms, an American Legion dance, the Catholic Youth Organization's national meeting in Denver, and, though they were underage, a regular gig at a local bar called the VIP. In the spring of 1968, they were playing in Denver when they heard gunfire in the distance—a mini-riot following the assassination of Martin Luther King Jr. Colorado Springs may have been a military town in the Vietnam era, but there was something innocent enough about the band that Paxton's Backstreet Carnival had complete buy-in from the older generation. General Moorman took some extra steps to make life easy for the band, scouting out the interstate in bad weather to make sure it was safe for the boys to drive to gigs. After school, Brian and his bandmates all walked to the Moormans' house, a big, private residence just around the corner from Air Academy High, where there was more room to rehearse. The band became such a fixture at the Academy that they played for visiting dignitaries. When Lucille Ball shot a two-part episode of her new show, *Here's Lucy,* at the Air Force Academy, she listened politely to Paxton's Backstreet Carnival and shook every band member's hand afterward. It was anyone's guess what Lucy made of them. And when Richard Nixon came to deliver a graduation address, five Secret Service agents in black suits interrupted the band's rehearsal, unable to believe there was a rock band rehearsing in the Air Force Academy superintendent's garage.

When Don and Mimi were away—taking the girls to Aspen or Santa Fe—Brian opened up the house for parties that seemed to draw the whole senior class, most of them smoking pot around Brian's younger brothers. Brian started taking LSD, too. But to Don and Mimi, Brian never seemed to be a problem. He was so talented! And so beautiful to look at. That Brian might have been suffering, unnoticed, just as young Donald had, never crossed their minds.

After graduation, Brian followed John to the music program in Boulder. He stayed a year before deciding that college was not for him. There was nothing keeping him local anymore—Paxton's Backstreet Carnival was no longer a going concern—so he made plans to go west with the hope of playing music and forming a new band. One of Brian's last local gigs made history, albeit not because of him. On June 10, 1971, he opened for Jethro Tull at Red Rocks, the concert amphitheater built into a natural shelf of outcroppings outside Denver. The show sold out quickly, and when more than a thousand fans showed up without tickets, the overflow crowd was diverted to a space a distance away. Some of those people started climbing a wall between that space and the amphitheater. Others charged the gate. That was when the police flew out in a helicopter and bombed the crowd with tear gas.

For decades, that show would live in infamy as the Riot at Red Rocks, Colorado's own miniature version of Altamont. Twenty-eight people, four of them police officers, were treated for injuries at the local hospital. Richard and Michael Galvin, then sixteen and eighteen, both remembered watching their rock star brother from a safe spot, away from the riot. Brian was up front, playing flute, as the police began cracking down—"just him and a guitar player," Michael said—not so far away that he couldn't smell the tear gas, but too focused on the music to register what was happening.

—⁓—

That same summer, 1971, Michael Galvin—the fifth son and the only one to proudly accept the label of hippie—was a newly minted high school graduate with no plan, and he could not have been more pleased about that. College was not on his agenda. Michael was not an ambitious person, but he somehow found a way to do what he wanted to do most of the time, and that was quite often enough for him. Altamont and the Manson Family and Kent State had all happened, but the bloom was not yet off the rose of the 1960s for Michael, nor was it for a lot of his friends. With the Vietnam War still raging, he wasn't so much a conscientious objector as someone who never got around to registering for the draft at all. Michael's plan, if you could call it that, was to ease his way in and out of any situation he found himself in, and see what happened next.

That summer was the start of Michael's separation from his family, the first step in becoming himself. First, he hitchhiked to Aspen, where everyone he met was in the middle of reading *The Prophet* by Kahlil Gibran and *The Teachings of Don Juan: A Yaqui Way of Knowledge* by Carlos Castaneda. Michael picked up both, and, in a way, never put either of them down again. It wasn't even what either author had to say, specifically, that reached something in him. It was the presentation of worldviews that had nothing to do with the austere Catholic upbringing he'd been forced to endure. These new ideas went down easily with pot and hash and LSD, but that was just part of the appeal.

From Aspen, Michael hitchhiked to Indiana with a friend, and then kept going east alone, hoping to get to New York in time for the Concert for Bangladesh at Madison Square Garden. He never made it. Instead, he stopped in Jerusalem, Pennsylvania, where he was arrested for taking a bath in a river. Michael spent eleven days in jail before a judge took pity on him and cut him loose. In Akron, Ohio, he was arrested again, this time for loitering. In front of the judge, he copped an attitude.

"Where are you from?" the judge asked.

"I'm from planet earth," Michael said.

He spent another few days in jail before finally deciding to call home.

"What can you do for me?" Michael asked his father.

"I'll send you a plane ticket," Don said. Somehow, as Michael remembered it, his father vouching for him was enough to get him out.

SETBACKS LIKE THIS didn't hit Michael that hard. "I think I was taking everything in stride," he'd remember later. Getting tossed in jail, sleeping in a park, or taking a bath in a river were all part of the same broader eye-opening adventure for him—a growing under-standing that reality was not necessarily what he'd once thought, that what he'd been brought up to believe may not be all there is.

The reality of being home, however, had never agreed with Michael. The 1960s, by his estimation, had somehow blown right past Hidden Valley Road. While other young people were off finding themselves, he and his brothers still had to dress alike, at least at church, wearing coats and ties on Sunday. Like the military, everyone was presumed to be the same, and everyone was expected to obey. If a Galvin son ever chose to question Mimi—which Michael made a regular habit of—she rarely settled for anything less than what she had first demanded.

Mimi, not Don, was the authority figure Michael was born to undermine. "My father was in the Air Force, but my mother was the *brains* behind the Air Force," Michael said. "He was gone," working two jobs and studying for his PhD. "She was our disciplinarian. So if we had to do hospital corners and make a bed perfect, that was because of her, not him." Mimi's lectures to the boys were epic, her ca-pacity to tune out any dissent practically endless. "You're not going to get the point across to her," Michael said; with Mimi, "it was always kind of a one-way street."

As a teenager, Michael's solution had been to not go home a lot. Hanging out with his friends, a joint in hand, he liked to think about Don being at Stanford in the late 1950s—around the same time as Ken Kesey, the countercultural icon who wrote *One Flew Over the Cuckoo's Nest* before leading a band of LSD-experimenting vaga-bonds across America. The idea of Colonel Don Galvin, the falcon man, dropping acid made them all laugh hysterically. At home, Mi-chael became bolder, rejecting the Galvin family's dress code, cutting the heels off of his Bass Weejuns so that they looked more like moc-casins. When Michael started showing up stoned, his father would sit down with him and talk, but not much changed.

Things got so bad that in the fall of 1968, when Michael was fifteen, Don and Mimi sent him to Jacksonville, Florida, to live with his uncle and aunt for the school year—a chance for him to get his head straight, learn self-reliance, and be one fewer problem for parents who, though Michael did not know it, were in the thick of dealing with Donald's issues. Michael took to Florida rather easily. His cousins, all a little younger than he was, found his Age of Aquarius affect fascinating. At his new high school, he had no trouble finding friends. He tried LSD for the first time on November 22, 1968; he remembered the date because it was the night Jimi Hendrix played Jacksonville. Michael went to the show with a new friend, Butch Trucks, who had just started a rock band with Duane Allman, and he spent much of the year hanging out at Butch's place. The following year, 1969, Butch and Duane's band became the Allman Brothers.

By then, Michael was back in Colorado, subject to the rules and regulations of the Galvin family for his last few years of high school. The only break in the monotony came in 1970, when Hidden Valley Road turned into a mental ward for Donald, lost and volatile after his divorce and hospitalization. Michael had no context for understanding Donald, and he was not terribly tolerant of Donald's chosen passion, the authoritarian Catholic Church. Michael started to lose his temper with his brother, and his parents weren't sure if the tension between them was Donald's fault or because Michael and Donald were too much alike. To a certain extent, their experience with Donald, along with Jim's delusional episodes, had shaken them awake. If two of their sons could lose their grip on reality, they were ready to believe that Michael might, too.

This was how the most formative moment thus far in Michael's young life came to pass, in the fall of 1971—not long after his high school graduation and his return home from his road trip and visits in the local jails of Pennsylvania and Ohio—when Don and Mimi sent him to Denver General Hospital, where he was held in the hospital's psychiatric ward on the top floor for observation.

Michael was prescribed Stelazine, an antipsychotic drug closely related to Thorazine. He wasn't there long, a week or so, before he decided that he was in the wrong place. He wasn't crazy—he was turning on, tuning in, and dropping out. He knew he did not belong there. So he left.

He slipped out of the hospital at his first opportunity, hitchhiked to a friend's house, and called his parents. "You can't make me go back there," he said. "I'm not coming home, either."

Don and Mimi were in a bind. Michael was eighteen, technically no longer theirs to control. They came back to him with a counter-proposal: How would he feel about going to California to visit his brother Brian?

Michael smiled.

AFTER BRIAN HAD left Colorado, his brothers still heard from him from time to time. Once, Richard got a letter in the mail with a joint inside, wrapped in red, white, and blue wrapping paper, along with a note that read, "Enjoy this from Jefferson Airplane."

It wasn't long, a few months, before his brothers learned that Brian had accomplished what he'd set out to do. He'd formed a new band with the name Bagshot Row, named for a street in the Shire near Bilbo Baggins's home in *The Hobbit*. This was exactly the sort of adventure Michael yearned for now. Nothing could have been more appealing to him than a chance to hang out in the Bay Area with a bunch of hippies and musicians—with his brother, the handsome, dark-haired prodigy, leading the charge.

When Michael arrived, he learned that not everything about Brian's new life was as advertised. Brian hadn't quite made it to the Bay. He and his bandmates were renting a house in Sacramento, an hour's drive from the coast. And Brian worked all day to pay the rent, leaving Michael on his own much of the time. What seemed like a perfect trip now was looking a little like a letdown. Bagshot Row was good, though—a rock-jazz-blues hybrid, featuring Brian as the band's flute soloist. Once again, Brian was the standout musician. But unlike his high school group, this band made original music, and planned to make records. Michael roadied for them a little, heaving the band's Hammond organ in and out of a van.

He wasn't there very long, just a month, before he got into trouble. Bored and alone one day, Michael decided that he wanted to go find the Pacific Ocean. He knew it had to be miles away, given this was Sacramento, but he had the time and he knew which way west was, and he thought if he could follow one of the canals or rivers, he'd get there. He spent the better part of a day walking before giving up and starting back to Brian's place. On the way, he cut through a trailer park and followed a dirt road. In the middle of the road, he noticed a garden hose connector. He picked it up, placed it on the step of the closest trailer, and knocked on the door. That got someone's attention.

The police picked him up just a few blocks from Brian's house. Michael heard one cop say the words "trespassing" and "attempted burglary." He was astonished. He didn't see how he'd done anything wrong. He figured he was being hassled for being a hippie. He got mad, and then he learned the police in Sacramento weren't as forgiving as that judge in Jerusalem, Pennsylvania.

In jail, Michael learned that attempted burglary was a felony charge. He'd never been in trouble with the law like that. While awaiting his court date, Michael tried to make friends. The guy in a neighboring cell taught him how to make toast with the Wonder Bread that came with meals: take your toilet paper and wind it up and light it with the matches you get for cigarettes, create a little campfire, and place your bread over it. Michael mastered that, and then he got caught for it. He got placed in solitary—a dark room where he was all alone. Until he was actually in there, Michael had no idea a place like that actually existed.

He was alone in there for days before he was offered a chance to talk to a doctor. Michael agreed, and the doctor he met arranged to move him to the hospital part of the jail. Michael had a room-mate and a TV now. That seemed like a move in the right direction. But next came another complete reversal of fortune: With no room available for him at Sacramento General Hospital, Michael was told that he was being transferred to Atascadero—California's notorious maximum-security mental hospital, holding two thousand inmates.

For the second time in the space of a year, Michael had been sent to a mental hospital—this time a mental hospital in a prison setting—and he could not have been more certain that there was nothing wrong with his brain. It had taken this moment—locked away with men who had killed their wives or their bankers or their kids—to finally shake him awake. This was not a lark, it was real life, happening to him.

Michael was told that he was only in Atascadero for observation, but nobody would tell him how long that was supposed to last. The uncertainty was as bad as anything else.

His father came to visit, but this time he couldn't do anything for him.

Brian came, too, but the best advice he could muster for his little brother was, "Life is about the journey, not the destination."

It was five months before the court let Michael plead guilty in exchange for time served. There was no way to explain this; Michael could only move forward, shake it off. His time in Atascadero wasn't without its diversions: Michael did meet one Yaqui Indian—a boxer who told a story about his brother fighting Sugar Ray Robinson— but the serendipity of that meeting was lost on him. He agreed with Brian: Life was about the journey. But some journeys, Michael decided, were better than others.

THERE WAS ONE thing Michael was sure about: He was not like Donald. He was not crazy. He would spend the rest of his life proving everyone wrong about that if he had to—his parents included. The problem here, he believed, was a labeling error. Not everyone who saw the world differently had schizophrenia. If that were true, every hippie would be crazy.

To help Michael make that argument, he had the entire 1960s ethos on his side. It seemed to a lot of people at that time that anyone who stood up and said no to authority, or rejected the military-capitalist superstructure, risked being labeled insane by those in power. By the 1970s, the public conversation about mental illness was no longer just about Freud and Thorazine. It was about seeing the diagnosis of mental illness as an instrument of conformity and power—just another way of clamping down on independent thought and freedom.

This was a countercultural position, but its roots ran back to the anti-psychiatry movement—a wave of therapists and others who, more than a decade earlier, had rejected traditional assumptions about insanity almost completely out of hand. In the 1950s, Jean-Paul Sartre had argued that delusions were just a radical way of embracing the world of imagination over "the existing mediocrity." In 1959, the iconoclastic Scottish psychiatrist R. D. Laing, influenced heavily by Sartre and other existentialists, made the case in *The Divided Self* that schizophrenia was an act of self-preservation by a wounded soul. Laing famously decried the "lobotomies and tranquilizers that place the bars of Bedlam and the locked doors *inside* the patient." He believed patients retreat inside their own mind as a way of playing possum, to preserve their autonomy; better to turn oneself into a stone, he once said, than to be turned into a stone by someone else. In 1961, the sociologist Erving Goffman published his book *Asylums,* in which he explored life in mental institutions and came away believing that the institution informed the illness of patients, not the other way around. That same year, the Finnish psychiatrist Martti Olavi Siirala wrote that people with schizophrenia were almost like prophets with special insight into our society's neuroses—our collective unconscious's shared mental illness. And again, that same year, the godfather of anti-psychiatry, Thomas Szasz, published his most famous book, *The Myth of Mental Illness,* in which he declared that insanity was a concept wielded by the powerful against the disenfranchised—a step in the ghettoization and dehumanization of a whole segment of society that thinks differently.

A year later, in 1962, anti-psychiatry crossed over into the mainstream with a juggernaut of a novel that treated the brutality of a state-run mental hospital as a metaphor for social control and authoritarian oppression. *One Flew Over the Cuckoo's Nest* was the story of Randle Patrick "Mac" McMurphy, a low-level criminal and free-spirited renegade who fights a war of wits inside of an insane asylum, only to be crushed by the malevolent forces of authority. Even before it became a movie, *Cuckoo's Nest* became one of the foundation myths for the counterculture, as romantic, in its way, and as powerful as *Easy Rider* and *Bonnie and Clyde*—a perfect way to explain the way the world was working right now and expose everything that had flattened out the culture of the previous generation.

Going back even further, of course, that idea of whatever society deems to be mental illness sharing the same wellspring as the creative, artistic impulse has been with us for centuries: the artist as iconoclast and truth-teller, the only sane one in an insane world. Even Frieda Fromm-Reichmann, in the years before her death in 1957, came to believe in a "secondary element" in the loneliness of some psychotics that makes them "more keen, sensitive, and fearless as observers." She wrote about the composers, artists, and writers with mental illness, suggesting their talents sprang from the difficulty they had with direct, conventional communication. Like a court jester, Fromm-Reichmann wrote, people with schizophrenia often tell uncomfortable truths that the rest of us would rather not hear. She was referencing the Cervantes novel *The Man of Glass,* about a village idiot who's treated tenderly by the people around him, as long as they can laugh off the painful truths he spouts as crazy delusions. But when the man recovers, the community prevents him from getting back on his feet, lest they suddenly have to take seriously everything he says.

By the late 1960s, the anti-psychiatry movement was no longer concerned just about the treatment of the mentally ill, or even about creativity or art—it was about politics, justice, and social change. In his 1967 book, *The Politics of Experience,* Laing argued that the insane people were sane all along—and that to call someone schizophrenic was, in essence, an oppressive act. "If the human race survives, future men will, I suspect, look back on our enlightened epic as a veritable age of Darkness," he wrote. "They will presumably be able to savor the irony of the situation with more amusement than we can extract from it. The laugh's on us. They will see what we call 'schizophrenia' was one of the forms in which, often through quite ordinary people, the light began to break through the cracks in our all-too-closed minds."

MICHAEL DECIDED THAT the only thing wrong with him was the repressive way he had been raised. "There was some kind of suppression," he would say. Michael believed conformity had corrosive power. He blamed practically all of his brothers' troubles on that. But even he had no idea how to help them. To him, they seemed trapped in prisons of their own making, and no one, not even he, had the keys to the locks.

In 1972, the authors Gilles Deleuze and Félix Guattari, in their Marx-meets-Freud mashup, *Anti-Oedipus: Capitalism and Schizophrenia,* called the family structure a metaphor for authoritarian society. Both the family and society, they wrote, kept their members under control, repressed their desires, and decided they were insane if they worked against the organizing principles of the larger group.

Schizophrenia had become a metaphor now. The theoreticians had left the idea of illness behind entirely, fixating completely on revolution. Families like the Galvins, meanwhile, were also left behind—collateral damage in a culture war—waiting for someone who actually knew how to help.

CHAPTER 14

1967
Dorado Beach, Puerto Rico

At a tropical hotel, under a blazing late-June sun, David Rosenthal—the researcher from the National Institute of Mental Health who had studied the Genain quadruplets and concluded that heredity and the environment must be working together—joined some of the most prominent thinkers in psychiatry at an academic summit about the continuing debate over nature and nurture and schizophrenia. Nothing like this had happened before, but a meeting seemed necessary now.

By the 1960s, the Thorazine revolution had raised the stakes in the debate. To those favoring genetics, or nature, the impact of neuroleptic drugs proved, at the very least, that schizophrenia was a biological process. But for the therapists on the nurture side, Thorazine and the like were just symptom suppressors—glorified tranquilizers—and there could be no substitute for probing the unconscious impulses that must have caused the disease. This conference, then, was a cautious attempt to break the impasse. While Rosenthal, as NIMH's chief researcher of schizophrenia, was one of the event's organizers, the psychotherapy camp was well represented, too—by, among others, Theodore Lidz, the Yale psychiatrist and pioneer in the study of family dynamics. The conference title, "The Transmission of Schizophrenia," was diplomatically worded; *transmission* was thought not to tip the scales in favor of one side or the other, the biologists or the talk therapists. Even the setting—Dorado Beach, Puerto Rico—seemed meant to ease tension and, just maybe, help build a lasting peace.

In the three years since he had published his book on the Genain family, Rosenthal had been approaching the nature-nurture question from a different angle. Toward the end of his work with the quadruplets, he had begun to see all too clearly the limitations of studying siblings who grew up in the same environment. Instead, he started to wonder what would happen to a child with a family history of schizophrenia if you raised that child away from her family environment. Who, in other words, would be more likely to develop schizophrenia: a genetically vulnerable child who grew up among her blood relatives, or a similar child who had been adopted and raised by people who did not share her genes? Now, in Dorado Beach, he was ready to announce the first of his findings. Here, it seemed to him, was proof that nature, not nurture, won the argument.

Rosenthal and Seymour Kety, NIMH's director of research, found a sample population for their study in Denmark—a nation that many genetics researchers had come to adore, thanks to that country's excellent medical record-keeping and willingness to contribute those records to scientific research. They were able to search the records for people who were adopted and then developed schizophrenia. Then they dug into the health records of the families who had adopted them, searching for correlations—to eliminate the possibility that an outsized number of the mentally ill adoptees happened to be adopted into families with a lot of mental illness. Finally, they compared their adoptees to a control group—schizophrenia patients who grew up in their own families. The end goal: See which scenario, nature or nurture, seemed to produce a greater incidence of schizophrenia.

It wasn't even close. At Dorado Beach, Rosenthal declared that biology, not proximity to people with a history of schizophrenia, appeared to explain nearly every single documented instance of the illness. Where you grew up, or the people who raised you, seemed to have nothing to do with it at all. On the whole, families with a history of schizophrenia seemed more than four times as likely as the rest of the population to pass along the condition to future generations—even if, as ever, the illness rarely passed straight from parent to child.

This conclusion spoke volumes about how the disease wandered and meandered through families—and that, alone, would have been stunning. But in their analysis of adoption cases, Rosenthal and Kety also found no evidence to support the opposing, "nurture" view—that schizophrenia could be transmitted from a mentally ill parent to an adopted child who did not share the adoptive family's genetic history. Schizophrenia, he concluded, simply can't be imposed or inflicted on someone who is not genetically predisposed to develop the condition.

Rosenthal thought he'd finally settled the argument—and, for good measure, discredited the idea that bad parenting created the disease. At the conference, he found at least one kindred spirit: a young geneticist named Irving Gottesman, who, with his coauthor, James Shields, had just published a study that reached a very similar conclusion. Their completed work, "A Polygenic Theory of Schizophrenia," argued that schizophrenia could be caused by not just one gene but a chorus of many genes, working in tandem with, or perhaps activated by, various environmental factors. Their proof involved twins, but with a twist: Instead of conceiving the illness as the handiwork of one dominant gene or two recessive genes, they proposed that there exists a "liability threshold" for genetic illnesses—a theoretical point beyond which some people might develop the illness. The causes that would collaborate to bring someone close to this threshold might be genetic or environmental—a family history of the disease and a traumatic childhood, perhaps. But without the critical mass of these factors, a person might live their entire life with a genetic legacy of schizophrenia and not become symptomatic.

Gottesman and Shields's theory became known as the "diathesis-stress hypothesis"—nature, activated by nurture. Decades later, their work would be seen as phenomenally prescient, the real beginning of the end of the great argument that had stretched back to Freud and Jung. Seen one way, the diathesis-stress hypothesis might even be interpreted as a compromise between the nature and nurture camps: If the theory held, then it seemed logical that Thorazine and other neuroleptic drugs, regardless of how they functioned, could only be one part of any lasting treatment of the illness.

But at Dorado Beach, the idea met with the usual resistance. Even one of Rosenthal's own colleagues from NIMH argued instead that a childhood spent in chaos or poverty could be one cause: The larger the city, new studies suggested, the more social class had a relationship to schizophrenia. But that same colleague acknowledged a causality question: Does poverty cause schizophrenia, or does congenital mental illness throw families into poverty?

The schizophrenogenic mother was back, too. A speaker from the University of Helsinki used his time to pillory mothers "embittered, aggressive and devoid of natural warmth" and "anxious and insecure, often with obsessive features." And yet the Helsinki therapist could not explain why, if the mother was to blame, some children of the same mother fall ill with schizophrenia while others do not. He had only his belief that bad mothering must be the answer.

Then came Theodore Lidz with his family dynamics explanation: A child can fail to mature adequately, he declared, if "he perceives very faulty nurturance in his first few years, or he is seriously traumatized." The Yale psychiatrist cited no data to support this position, just his personal work with families affected by schizophrenia.

A week went by like this until, on July 1, the conference's last day, it fell on Rosenthal, the organizer, to sum up the state of the field. He treaded lightly, opening with a joke. The heredity-environment controversy in schizophrenia, he said, reminded him of a "white-shirted French duel," in which the duelers "managed to avoid each other so thoroughly that they never exposed themselves even to the danger of catching cold." Remaining diplomatic, Rosenthal said that he saw it as a positive sign that everyone was able to come together at all. "This week we have been able to sit here day after day and listen to people expounding ideas both compatible and contrary to our own," he said, "and far from catching any dread affliction, the only thing we have caught, I hope, is the spirit of earnest concern about the other man's data and opinions."

There would be no real reconciliation anytime soon. Three years later, the chief of the family studies section at NIMH, David Reiss, also a participant at Dorado Beach, would still be referring to the geneticists and the environmentalists as "warring camps." Families like the Galvins, meanwhile, continued to live at the mercy of a mental health profession still caught up in a debate that came nowhere close to helping them. But there was a good reason for this impasse, one that Rosenthal acknowledged in his closing remarks—a mystery that would take another generation to even start to be solved.

The good news, Rosenthal said, was that "all the reasonable doubts that had been raised in past years have now been answered, and the case for heredity has held up convincingly." This conference, he predicted, "could be remembered as the time when it was definitely and openly agreed by our foremost students of family interaction that heredity is implicated in the development of schizophrenia."

But that concession only raised a more puzzling question. "In the strictest sense, it is not schizophrenia that is inherited," he said. "It is clear that not everybody who harbors the genes develops schizophrenia." Schizophrenia was definitely genetic, but not always passed down. And so they all were still left wondering: How could this be?

"The genes that are implicated," Rosenthal said, "produce an effect whose nature we have not yet been able to fathom."

DON
MIMI
DONALD
JIM
JOHN
BRIAN
MICHAEL
RICHARD
JOE
MARK
MATT
PETER
MARGARET
MARY

CHAPTER 15

Nothing may have been more important to Mimi than a flawless Thanksgiving. She spent all day on the meal, and beforehand she usually made a gingerbread house in time for it to be on display. In recent years, as Mimi had been forced to look past the food fights and dishtowel whippings between the brothers, each November still filled her with hope, offering one more chance for a beautiful experience.

This year, 1972, Joseph and the other three hockey boys were all still at home with the two girls. Donald, too, was home from Pueblo. As the day wore on, Jim and Kathy and baby Jimmy joined them, along with Brian and Michael and Richard. Only John was away, with his wife Nancy's family. With this many Galvins in one place, the chances for an explosion were high. The sparring started early and continued up until mealtime—the boys sniping at one another over who took how much to eat, who did their share of cleaning up, who was a pansy, who was an asshole.

You took too much!
What are you going to do about it?
You didn't leave me any
Too bad for you
Move over

You stink
You suck
Fuck you
You asshole
It's not my turn to do the dishes
You never help around here
Pansy
You are such a girl
Take it outside
Grow up

Margaret braced herself. On Thanksgiving, it fell on her, now ten years old, to iron the linens and place the silver and the napkins on the table. These chores kept her close to her mother and away from the boys. Keeping with the family tradition, there were assigned seats. Don, the patriarch, was at the east end of the table, with Donald to his immediate right, where he could be closely monitored. Mimi's place was on the table's north side, with a view out the window, with chess-playing Mark and introspective Joe nearby, and rebellious Peter closest of all, so that she could keep an eye on him. Margaret always sat at an end because she was left-handed, and little Mary, still just seven, not far away. Matt sat across from them, near Jim and Kathy. But they weren't seated yet this year when the worst happened.

Jim and Donald were more at odds than ever. They fought every time they were in the same room now. Jim looked at Donald and saw a weakened foe, someone he could finally defeat; he also might have seen an unwelcome image of himself, suffering from delusions just as he was. Either way, Donald had to be expunged, and Jim had to be the one to do it. Donald, meanwhile, looked at Jim and saw a pest who never seemed to go away. He'd been humiliated enough—by a wife who would not agree to stay married to him, by brothers who did not obey him the way he'd once hoped. For Jim to walk through the door and assume that he was in charge was, for Donald, the final insult.

So they fought—wrestling, like in the old days, in the living room, the usual spot. Donald used to have the advantage, but not anymore; Donald had been in the hospital, and was weakened by neuroleptic drugs. They seemed evenly matched now. As someone got little Mary safely out of bounds, the fight escalated.

It wasn't long before they could not be confined to one room.

The living room at Hidden Valley Road opened out to the dining room. If you wanted to take a fight out into the backyard, you would have to cross through the dining room to get there. On this Thanksgiving, the brothers started to move in that direction. The only thing in their way was the table.

Donald ran to the far side of the dining room. He lifted up the table, with Jim on the other side, coming closer. In Margaret's memory, he tipped the table over onto its side, and everything on it came crashing down onto the floor. In Mark's, Donald actually picked up the whole table and threw it at Jim. In either case, Mimi's perfect Thanksgiving was destroyed.

Mimi looked at the house now, at the table keeled over, at the plates and silver everywhere, the linens crumpled in a heap. There may have been no better, more precise manifestation of her deepest fears than this, no clearer way of illustrating the way she felt just then—that every good thing she had done, all the work, all the attention to detail and love, yes, love, for her family was in pieces. There was no sugar-coating this. Her mother, Billy, if she'd been there to see it, would have known without a doubt how bad things were—how profoundly Mimi had failed. Anyone would.

She turned her back on everyone and walked back into the kitchen. That was when everyone heard another noise, softer this time: the gingerbread house, being smashed to bits by the woman who had made it.

"You boys don't deserve this," Mimi said, in tears.

Running almost as a dividing line between the Galvin and Skarke properties at the end of Hidden Valley Road was a small trail that had gone unused, seemingly, for years. One day, the Skarkes bought a Honda 90 minibike. Carolyn Skarke, who was about Margaret's age, would ride up the trail between her house and the Galvins' house to visit one of her other friends. The trail was technically on the Skarkes' property, but no one had ever talked much about it until Carolyn started to use the motorbike on it.

One day, Carolyn was riding down the hill on the trail and had almost made it to the bottom and home when, by a stroke of luck, she noticed a cable, thin as a wire, strung across the path, blocking access to the Hidden Valley Road cul-de-sac. She was able to veer away from the cable at the last second, just before she got clotheslined. Frightened nearly to death, Carolyn told her mother, who, as soon as she determined what had happened, marched out of her house, past the trail, and toward the Galvin house, searching for Mimi.

Carolyn remembered watching the two women, who had always been civil, standing outside on that little road, facing off like a fuming baseball manager and a stubborn umpire.

"Why did you do this?" her mother shouted.

"I don't like the noise," Mimi said.

That was all that Carolyn's mother could take.

"We put up with all the *sheriff's cars* coming to your house? *And you don't like a Honda 90??*"

Everyone knew something was happening at the Galvins'. Their closest neighbors pulled out of their driveways with care because they knew there was a good chance that Donald would be loitering along the road, offering prayers to everyone who drove by. The younger boys were becoming well known, too. Matt got caught taking things from a neighbor's house when the family came home in time to see him. And Peter had developed a haunted, menacing look that some of the girls would comment on. Soon enough, it was more than his looks that worried them. Once Peter stuffed a girl's face in the snow and kept it there until she couldn't breathe, then insisted it had all been just a joke.

Practically no one visited anymore. The Hefley kids weren't allowed to come and play there. And anytime anything happened in the neighborhood—if someone's mailbox was vandalized, or a house broken into—plenty of people were ready to blame it on the Galvins.

Mimi made a practice of denying everything. "My boys would not do anything like that." No one believed her. She was silently drowning, left alone to manage a situation for which she had no tools and no training and no natural aptitude. Both she and Don had taken to falconry because it made sense. Their children did not make sense. They had tried to instill procedures and routines to train their children. But children aren't falcons.

What did change was that Mimi became embittered. If a child

stepped out of line now, she was no longer the happy warrior—she was the angry general. Her frequent refrain to Michael or Matt or Richard or Peter, whenever they disobeyed, was, "You're just like Donald." She might not have realized how lethal that phrase could be. To accuse the boys of being like Gookoid was probably the worst thing she could say—a reminder that they shared blood with this man, this stranger, who was turning their home into an unbearable place, who was ruining all of their lives.

THERE WERE STRETCHES, a week or a month, when Donald would show flashes of lucidity, and even hold on to jobs—dog catcher, land salesman, construction worker. In 1971, he was switched to a related antipsychotic drug, Stelazine, and his outlook dramatically changed. "He realized in the space of one weekend that much of his spiritual musing was simply his imagination and not reality," a Pueblo psychiatrist named Louis Nemser wrote. "He described how his desire to build a church was related to him being more like his ex-wife's father—and the magical wish that if [Donald] were more like him, she would take him back."

His progress lasted several months, until, in April 1972, after yet another failed mission to see Jean in Oregon, Donald visited a priest at the Catholic chancery to talk about his marriage. The priest told Donald in no uncertain terms that his union with Jean was now null and void in the eyes of the Church—a pronouncement that sent him straight back to Pueblo, where, Dr. Nemser sympathetically reported, "it seemed like his tears would never stop."

The psychiatrist, taking a page from Freud and Frieda Fromm-Reichmann, seemed to believe that Donald somehow had the ability to climb out of this state on his own—"It seemed that Donald had chosen to become psychotic again," Nemser wrote—and so he decided the staff should do what they could to help Donald choose wisely. They went out of their way to empathize with him, holding his hand and telling him how sad they were for him. The strategy did have some effect on Donald. "He began to express himself more freely about Jean," he wrote, "saying that he still really cared for her and hoped that she might one day contact him, but that he refused to contact her anymore since every time he did this, it ended in disaster for himself."

Donald worked his way toward being discharged again, finding a job lead while out on a one-day furlough. He would be a vacuum cleaner salesman—good money, flexible hours, everything he could reasonably want. He got out on May 2, 1972, and the cycle started again. This time his behavior became so menacing—threatening both of his parents' lives, they said—that Don and Mimi petitioned a local court to order Donald back into Pueblo in August. When the staff decided to place him in seclusion, he grabbed the keys from the door, pushed the attendant into the seclusion room, and locked him in. Donald didn't try to escape, however. He just sat there outside the room, saying he wanted to teach the attendant a lesson. The doctors gave him a high dose of Thorazine and a smaller dose of Stelazine. Gradually, he climbed out of psychosis again, and he was discharged on August 28 with a guarded prognosis.

The following spring, 1973, Donald was admitted to Oregon State Hospital after trying once more to see Jean. His intake notes described him as "very uncooperative and unmanageable" and "both confused and disoriented"—saying, for example, that he had no memory of ever being at this hospital, despite this being his third time there.

BETWEEN HOSPITAL STAYS, Donald was home in time for another family wedding, this one a less pleasant affair than John's a year earlier. Richard, son number six, had been the Galvin family's schemer—ambitious and gutsy, entrepreneurial, and more than a little willing to bend the rules to get what he wanted. As a freshman at Air Academy High, he found a way to sneak into the school commissary by applying chewing gum to the lock of a door. For months, he and his friend stole jeans and food and anything else they could get their hands on. Once he was caught, he was suspended for a year and forced to go to another high school, which had infuriated his father. "You're going to mess up if you keep going this route," Don would say.

In 1972, Richard was back at Air Academy High, completing his senior year, when he scored the winning goal in the state hockey championship. After the game, a girl from the opposing team's cheerleading squad asked Richard if he was going to the victory party. He was, and she got pregnant that night.

Their wedding took place a few months later, somewhat under duress, at the Garden of the Gods, the geological marvel in Colorado Springs. Richard was high on mushrooms during the ceremony. His friend Dustin played "The Times They Are a-Changin'" on the guitar. Still, all seemed to be going smoothly until a voice cut through everything. Donald had climbed to the top of the rocks. Now he was shouting: *"I'm not allowing this marriage! This marriage is not in the truth of God!"*

Jim and Don subdued Donald, and the ceremony went on.

IN MAY 1973, Don and Mimi agreed to take Donald back from Pueblo one more time. This time he lasted four months.

On September 1, a member of the sheriff's office brought Donald back to the state hospital on another "confine and treat" order, requested by his parents. Donald told the intake staff that he'd taken his Stelazine, as instructed, but when he asked Mimi for some Benadryl, she refused to give it to him; she was worried it would make him too sleepy to drive safely. That was when Donald lost control, grabbed her by the throat, and began to shake her. Yet again, it took a few of the other boys to keep Donald from strangling his own mother.

Back in Pueblo, Donald asked to see a Catholic priest. "He denies hallucinations or delusional paranoid feelings," the hospital staff conference notes read. "The only problems he admits to are emotional problems that he encounters with his family, which apparently includes some physical fighting." His prognosis, once again, was designated "guarded."

Within a few days, Donald's troubles would be the last thing anyone in the family was thinking about.

DON
MIMI

DONALD
JIM
JOHN
BRIAN
MICHAEL
RICHARD
JOE
MARK
MATT
PETER
MARGARET
MARY

CHAPTER 16

In a home that now rarely knew a moment's peace, an appearance by Brian, riding in from California for a family visit, was a welcome balm, a break from the pathos and a shot of electricity—the rock 'n' roll star, or at least the Galvin family's version of one, returning home. When he turned up with a girlfriend, that got everyone's attention. The couple communed with everyone in the Galvin living room, playing reel-to-reel tapes Brian had brought of his band, Bagshot Row. He brought his guitar and played along with his brothers, and the air pressure of the place changed completely. Mimi even let the couple sleep in a room together downstairs—a special dispensation that spoke to Brian's elevated status in the family.

Lorelei Smith, or Noni to her friends, was a native Californian—bright, cheerful, and no-nonsense, with sun-kissed blond hair and a friendly smile. She was three years younger than Brian, and her childhood had been far more luxurious than his. The walls of Noni's childhood bedroom in Lodi, a small town outside of Sacramento, were covered with ribbons from horse shows. But there was more than enough heartache and strife in Noni's life to interest Brian, who had always seemed drawn to the darker aspects of the human condition. Noni was barely a teenager when her mother died from a combination of pills and alcohol. Her father, a well-known pediatrician in town, married a woman from the horse show set who was less than ten years older than Noni was. Noni never lived with her father again. She spent three years at boarding school, and her senior year at her sister's house in Lodi, so that she could finish at the local high school. By the time she and Brian were together, Noni had found work at a veterinarian's office in Lodi while taking classes at a business college.

Nearly a half century on, few people are left who remember Noni. Her father, mother, stepmother, and sister have all passed away. Her sister's former husband remembers her as a happy girl, likable and charming. She had floated through Lodi High School for just a year, not long enough to make a lasting impression. The only member of the next generation who was alive at the same time as Noni is a nephew, the son of her sister—now an adult, nearly twice the age Noni was in 1973. All that nephew has is the awareness that once there was a girl named Noni whose boyfriend shot and killed her—and that after that, no one in his family ever was the same.

BEFORE HE'D LEFT for California, Brian had been given to hip-pieish philosophical musings. He had talked about death, but not in a grim or fatalistic way—more as if it were a state of mind, a crossing over to another dimension. "To him, it wasn't ending," said John, the third son, who roomed with Brian for a year in the music program at CU Boulder. "It was just going somewhere else. He'd always talk to me about going over to the other side."

To John, there didn't seem anything too urgent or dangerous about the way he was talking. "It was the times," John said. "The psychedelic times that we lived in." Some of this could have been fanned by drugs; no Galvin brother dropped more acid than Brian. But there was a darkness to Brian that never seemed to concern his brothers, either because they couldn't see it, or because they didn't want to see it, or because they found it romantic.

On the afternoon of Friday, September 7, 1973, the Lodi police department received a phone call from Noni's boss's wife at the Cherokee Veterinary Hospital, concerned that Noni had gone home for lunch at noon and had never come back. An employee missing for an hour or two would hardly seem to rise to the level of a police matter—

unless there was something happening with Noni that everyone at the office knew about, something that made her vulnerable.

Noni and Brian had broken up a month or so earlier. They had been arguing ever since. And now, Noni was living alone.

The first officer to arrive at 404 ½ Walnut Street found the apartment door open. He walked inside and found the young couple on the floor, a .22 caliber rifle beside them. Noni's face was covered in blood. She had been shot in the face. Brian had a gunshot wound to his head—a wound that the police on the scene determined to be self-inflicted.

THE YOUNGEST CHILDREN—Peter, Margaret, Mary—awoke to the sound of their mother sobbing.

Downstairs, Mimi was lighting candles on the kitchen table, and Mark was trying to calm her down. Don was on the phone, making arrangements, pulling their brother Donald out of Pueblo on a temporary pass so that he could attend his brother's funeral.

The official explanation, at least for the little ones, was a bicycle accident. Margaret was eleven and Mary almost eight, too young to be told that Brian had shot and killed his girlfriend, and then turned the rifle on himself. Many of the others didn't get the full story, either. Some believed the couple had been the victims of a robbery gone wrong. They most likely would not have thought that, had they been told what the police had learned—that Brian had bought the murder weapon from a local gun shop just a day earlier. What happened in Lodi seemed premeditated.

Years later, others in the family entertained other theories—that Brian and Noni had a suicide pact, or had taken LSD together. But what only Mimi and Don knew, and told no one for many years, was that sometime before his death, Brian had been prescribed Navane, an antipsychotic. There is no known record of the diagnosis that called for that prescription—mania, or depressive psychosis, or trauma-induced psychosis, or a psychotic break triggered by the habitual use of psychedelic drugs. The other children never learned when their parents first knew about this. But both Don and Mimi must have understood that one of the conditions Navane treats is schizophrenia. The thought of another insane son—their amazing Brian, of all people—was so devastating to them, they kept his prescription secret for decades.

MICHAEL WAS NUMB. He had been on his way to California, but had stopped in L.A., thinking he'd get around to seeing Brian up north some time later. Now all he could think was that Brian needed someone to throw a wrench in whatever it was that had been set into motion—and that he hadn't been there to help. Now he was asked to help again: His father recruited Michael to come with him to California to get Brian's body and find something to do with all of Brian's belongings. They met with the police, but as an officer explained to him and his father what they thought had happened, Michael couldn't handle it. He tuned out, refusing to hear anything more, about a second after he heard the words "murder-suicide."

Even without knowing about Brian's prescription, the younger boys connected what had happened to what was happening to their older brothers: first Donald, then Jim, and now Brian. John's wife, Nancy, was the first to say out loud what everyone else had to be thinking—that what was happening to the Galvin boys had to be contagious. She and John left Colorado for Idaho, where they both found jobs as music teachers. The other sons started to drift away. Joe, the seventh son and the oldest of the four hockey boys, moved to Denver to work for an airline as soon as he graduated high school. Mark, the next in line, graduated a year later and headed off to CU Boulder.

After a brief furlough for his brother's funeral, Donald returned to Pueblo—"quite intense about his religion," the staff reported that year, "extremely controlled" in affect, again with an "underlying hostility close to the surface." He stayed for more than five months, returning home in February 1974 with some new medications: Prolixin, an antipsychotic alternative to Thorazine; and Kemadrin, a Parkinson's drug often prescribed to temper the side effects of neuroleptic drugs. Not counting Donald, Don and Mimi had just their four youngest children left at home: Matt, Peter, Margaret, and Mary.

DON
MIMI

DONALD
JIM
JOHN
MICHAEL
RICHARD
JOE
MARK
MATT
PETER
MARGARET
MARY

CHAPTER 17

Don had spent years building distance between himself and his children. Even once they started getting sick, he kept working, out of necessity but also in such a way that it removed him from the day-to-day dramas, just as he'd always been. Two months after Brian's death, he acquired an additional professional title, beyond his role at the Federation: president of the newly formed Rocky Mountain Arts and Humanities Foundation.

But what had happened to Brian proved impossible for any of them to move past, and while Mimi searched for ways to keep busy with the children who remained at home, Don internalized it all. Early one morning in June 1975, Don was getting ready to leave the house to take Peter to an early morning hockey practice when he collapsed to the floor.

The stroke hospitalized Don for six months. He was paralyzed on the right side of his body and seemed completely without short-term memory. As he regained control over his body, he still couldn't remember anyone's names, or much of what had happened in his life after World War II.

Don reluctantly announced his retirement. The farewell letter from the Federation was courteous, if a little cool. "In light of your recent stroke," wrote the governor in charge, for whom Don had done all the grunt work, "I think your decision to seek a job which gives you greater control over time, travel, and responsibility, is a wise and sound decision."

After years of leaving his wife to take care of the children, Don now needed Mimi to take care of him. Don had always thought that the sick boys ought to leave and get treatment outside the home. "God helps those who help themselves," he would say; if the boys were unwilling, there was nothing else anyone could do. But now Mimi had her way with no protest from Don—in part because Don, in his weakened condition, had lost the authority to make decisions; and in part because they had let Brian go, and look what had happened to him.

All of Don's old arguments—that Mimi had been babying the boys; that he believed in the school of hard knocks; that those self-help books he gave the boys were all about pulling yourself up by your bootstraps—would never work again. Now that the worst had happened, Mimi would never give up on another one of her sick children.

—∾—

As the youngest of ten brothers, fourteen-year-old Peter seemed to have so much authority weighing over him that he chose to disregard it all, starting arguments and defying orders every chance he could. He was so rebellious—oppositional defiant disorder might have been his diagnosis, a generation or two later—that Mimi got into the habit of calling him a "punk," picking apart anything he did that was out of step with what was expected. If this seemed a little harsh, Mimi felt justified: Just when it seemed as if things couldn't get any harder for the family, Peter seemed to be going out of his way to make things even worse. But what bothered her most, of course, was the feeling that if Peter veered too far off course, he would go the same way as Donald and Jim and Brian.

Peter might always have been contrary, but his father's stroke—which he'd been present for, had witnessed, helplessly, from just a few feet away—seemed to shake loose whatever self-regulating mechanism he once had. He was caught stealing things and even setting a small fire. Then came the morning in his ninth grade algebra class, not long after Don's stroke, when he started talking gibberish to the students around him. When Peter's teacher tried to get him to stop, he wandered over to her, sat on the edge of her desk, and kept talking. After she got Peter back into his seat, the principal and dean of students came to the classroom. They brought along a third man, a gym teacher, in case Peter got violent.

Peter was admitted to Penrose Hospital in Colorado Springs, but only briefly—just long enough for the doctors to stabilize him. Once he was home, Mimi, her hands full with her husband's hospitalization, decided to send him off to hockey camp as scheduled. It was there that Peter fell apart completely, wetting his bed, spitting on the floor, hitting the other campers. He left the camp for Brady Hospital, a private psychiatric clinic in Colorado Springs, where the doctors prevented anyone from visiting Peter for weeks.

In early September, Mimi finally visited and saw Peter wearing only underpants, strapped to a bed with no sheets on it. The whole room reeked of urine. Mimi pulled him out right away. Before leaving, Peter was prescribed a small dose of Compazine, a drug usually used for nausea and vomiting.

Mimi was running out of options. The state mental hospital in Pueblo that treated Donald seemed like too much, too extreme, for a boy of his age. So Peter's next stop, late on a Saturday night in September 1975, was the University of Colorado Hospital in Denver. Peter was in the waiting room for so long that he started urinating. Once admitted, his speech was too slurred to be understood.

"It was sad to note that when the patient did become more provocative," the doctor wrote, "his family thought this was his normal level of functioning."

IT WASN'T LOST on the doctor that both Mimi and Don, when he was well enough to visit, referred to Peter as the latest of their sons to have lost his mind. Before long, the staff of the hospital learned about the others.

They learned about Donald, and a troubling dynamic he seemed locked in with Peter—how the more strangely Donald behaved, the more heat Peter seemed to catch for it at school, and the more Peter came to resent Donald at home. "It is easy to be number one," Peter used to say, "but not everyone can be number ten."

They learned about Jim, who happened to have been admitted

as a patient in the adult psychiatry ward of the same hospital, after experiencing what the staff identified as "an acute schizophrenic state highlighted by severe paranoid ideation."

They learned about Brian and the murder-suicide. And they saw for themselves that something was off about Joe, the introspective seventh son. When Joe visited Peter on the ward, one doctor wrote, he "was able to tell the patient's therapist that at times in the past he has had symptomatology similar to Peter's."

Here was what appeared to be case number five, coming down the pike. There was nothing in the medical file that suggested there was anything to be done about Joe other than to watch carefully for signs of the same psychosis that took command of the others. All this confirmed everything Don and Mimi feared. Something was happening to all the boys, one by one—first Donald and Jim, then Brian, and now Peter and soon maybe Joe—and they had no idea how to stop it, or even if it could be stopped.

Casting about for clues, Mimi and Don wondered if each brother had been set off by some sort of heartbreak: Donald's and Jim's marital woes, Brian's breakup with Noni, Peter seeing his father collapse with a stroke. Mimi also searched for something in their family histories—a precursor in some distant relative that could have warned them about this. Don's mother was depressed once, and so was Don after the war. What about that emotional episode Don had in Canada? Wouldn't you call that a breakdown? Was Don the carrier of a plague that all the boys were, sooner or later, destined to catch?

Or maybe drugs were to blame. Where the boys once listened to the Metropolitan Opera, now they blasted Cream and Jimi Hendrix. Brian, Michael, and Richard had all been into LSD—mild-mannered Joe, too. Chess-prodigy Mark was into black beauties and other uppers. Even Mary smoked pot at age five, thanks to Peter and Matt, who probably scammed it from one of the older brothers. Don and Mimi had noticed at least some of what was going on at the time, but they found themselves with very little power over so many boys. They never could have predicted how drugs would be everywhere, suddenly—at least not for their own exceptional children.

Now, for them, the counterculture became suspect. Could what was happening to their boys be, in some way, just another aspect of the volatile, rebellious times they were living in?

But the doctors who saw Peter had another theory.

THE NOTES FROM Peter's hospital stay in 1975 are extraordinarily tough on Mimi. One doctor wrote that she was "unwilling or unable to hear unpleasant news," and very adept at giving Peter "mixed and double messages"—a reference, it would seem, to the double-bind theory of bad mothering—and "successfully thwarted him from stating conflicting areas."

In therapy sessions where Mimi was present, a doctor noted, Peter would try to bring up his hallucinations and fears, but Mimi would not allow such talk to continue. "It seemed apparent that this role has been played by mother with the other sons as well," the doctor wrote.

At the same time, there was no question that both Mimi and Don were worried about their son, and Mimi was inarguably a source of comfort to him. "At times during the family meeting," the doctor wrote, "the patient would rest his head on his mother's chest and would show a smile, which made one think of a contented infant." To the doctor, at least, this dynamic—omnipotent mother and dependent baby—"was most comfortable for mother and her children to fall into."

At one meeting that Mimi would never forget, she and Don, sitting at a large table, flanked by doctors, found themselves directly on the receiving end of the schizophrenogenic mother theory. Everything they were telling Mimi added up to her being the prime mover in Peter's—and by extension all the others'—mental breakdowns. They both were stunned. Mimi was first appalled, then horrified, and finally defensive.

She resolved never to let the university doctors near her sons again. From then on, it would be Pueblo or nowhere at all.

ONCE, MIMI HAD thrived on structure and order, but now life offered her nothing close to that. With each new sick boy, she became more of a prisoner—confined by secrets, paralyzed by the power that the stigma of mental illness held over her.

Now the pretense of normalcy was a luxury. All the anguish

she'd tried to keep secret for so many years, she could not wish away anymore.

Exactly what, again, had brought Mimi Galvin to this moment? One son dead, a murderer; her husband laid low by a stroke and incapacitated; two profoundly ill sons at home, with no one to care for them but her. Only one more boy, Matt, sixteen, remained with the girls, thirteen-year-old Margaret and ten-year-old Mary. Caring for them all, and whoever might get sick next, was too much for Mimi, or for anyone.

It was at this moment that, one evening over the Christmas holiday in 1975, the phone rang in the Galvin kitchen. Mimi answered. It was Nancy Gary, Don and Mimi's Federation friend. The oil baron's wife.

Nancy in no way could have been the person Mimi most wanted to talk to at a time like this. Even hearing Nancy's no-nonsense, brasstacks voice on the phone, Mimi felt almost like she was hearing an echo of her old life, calling out to her, taunting her. Jetting to Salt Lake or Santa Fe on Nancy and Sam's private plane seemed like a life she would never lead again—a future destined now for anyone else but her.

But Nancy turned out to be the right person at the right time. She asked Mimi how she was doing, and for the first time, Mimi let her guard down. She did something she never imagined herself doing: She broke out into sobs on the telephone to a woman she only barely considered a friend.

Nancy was not an emotionally demonstrative person. But if there was one thing she was good at, it was using her husband's fortune to make problems disappear.

"You've got to get those girls out of there," Nancy said. And then, as quickly and easily as if she were ordering room service, she added: "Send me Margaret."

MARY GALVIN KNEW she was not supposed to show her feelings. After everything that had happened, her mother would verge on hysteria anytime either of her little girls lost control.

But as her sister, Margaret—the only person close to an ally she had in this house—packed up to leave, Mary cried harder and louder than she'd ever remembered. She was so visibly distraught that her parents feared she would make a scene when they dropped Margaret at the Garys'. They wouldn't even let her ride in the car.

Instead, on a January day in 1976 that is forever seared into her memory, Mary, just ten years old, stood at the front door on Hidden Valley Road, shrieking uncontrollably as they drove her thirteen-year-old sister away, leaving her behind with Donald and Peter—and a third brother, Jim, waiting in the wings, offering what she was told was a refuge, but even then knew in her heart was not—and feeling as abandoned and adrift and helpless as she'd ever felt in her young life.

Part Two

CHAPTER 18

1975
National Institute of Mental Health,
Washington, D.C.

There were many days when Lynn DeLisi felt she was in the wrong place at the wrong time—that she didn't belong in science, and that she'd been foolish to think she ever did. But the worst might have been the day she was told she might be driving her own children crazy.

This prognosis came her way from, of all people, a child psychiatrist at the National Institute of Mental Health, supposedly the vanguard of American psychiatric research. He said it in passing, in the middle of a lecture. DeLisi was one of a few women in the room, a first-year psychiatry resident at St. Elizabeths Hospital in Washington, D.C. And she was the only mother—of two toddlers who, at that precise moment, were at home being cared for by a sitter. Like Yale's Theodore Lidz, the family dynamics specialist, this psychiatrist seemed to believe in a connection between mental illness—specifically, schizophrenia—and the rise of the working woman. Mothers, the psychiatrist told the residents, ought to devote the first two years of their children's lives to being with and caring for them at all times.

DeLisi couldn't help but feel singled out. Of all the mothers in DeLisi's neighborhood in suburban Annandale, Virginia, she was the only one who left her children with a nanny while she went to work, in Washington. Her husband worked, too, of course, but it fell on her to tailor the demands of her residency around parenthood: To avoid night calls, she had made a special arrangement to make up the hours by extending her residency longer than the normal allotted time.

While the other first-years remained silent, DeLisi started arguing, demanding some sort of proof. "Where is the evidence?" she said. "I want to see the data."

But this psychiatrist had no data. He was citing not studies but Freud.

For weeks afterward, DeLisi could not stop thinking about how what he'd said with such certainty was informed not by experimentation and verification, but by anecdote and bias. What happened that day would color everything about DeLisi's career in the years to come. In an era when schizophrenia treatment was torn between two approaches—psychotherapy or psychoactive medications—DeLisi was drawn to a third way: the search for a verifiable neurological cause of the disease.

SHE HAD WANTED to be a doctor ever since she was a little girl in the suburbs of New Jersey. Her father, an electrical engineer, supported her dream and encouraged her; he would be the last man to do that for a while. She first thought of studying the brain and its relationship with mental illness at the University of Wisconsin, reading whatever she could about the neurological effects of hallucinogenic drugs. But her timing was not ideal. She graduated in 1966, when the Vietnam War was motivating many of her male classmates to apply to medical school to get deferments. Women who applied alongside those men were going in with an automatic disadvantage: Why would a medical school give a slot to a woman, when every man they turned down might be sent off to war?

Lynn struggled to find a work-around. She took a year off after college and found full-time work as a research assistant at Columbia University, and took graduate classes in biology at night at New York University. In the science library, she met the man she would marry, a graduate student named Charles DeLisi. Before their wedding, she enrolled in medical school at the only place that would take her: the Woman's Medical College of Pennsylvania in Philadelphia. After her first year, she applied to transfer to schools in New York, where Charles was still in graduate school. One interviewer asked if her family was more important to her than her career; another asked if she planned on using birth control. No one would take her.

Even her husband had expected her to quit medical school and switch over to a less demanding graduate program. But she stayed in the program with help from the dean—a woman who made a practice of championing young women in the profession—who arranged for DeLisi to take her second year of classes at NYU medical school. The next year, when DeLisi's husband got a postdoc at Yale, they moved to New Haven; DeLisi commuted by train all the way back to classes at the Woman's Medical College in Philadelphia. When she got pregnant with her first child, her dean stepped in to help her again, arranging for her to take classes at Yale for her entire final year.

DeLisi graduated medical school in 1972. Once again, they moved for her husband's work, this time to New Mexico, where Lynn started a general practice, treating poor migrant workers. She gave birth to their second child there, and when her husband had a job offer in Washington, she applied for residencies there. "I was interested in schizophrenia because it was a real disease of the brain," she remembered. "It wasn't, you know, just everyday anxiety. It was a real neurological disease." Being close to NIMH, DeLisi thought, would put her alongside people who felt the same way.

It took time for her to find those people. While St. Elizabeths had wards full of schizophrenia patients, the study of schizophrenia as a physical ailment was not in fashion, at least among the supervisors of her residency program. One problem was practical: The patients never seemed to get better. Better to spend your career working on depression or eating or anxiety disorders or bipolar illness—something with even a glimmer of hope, that sometimes responded to the traditional cure of talk therapy.

Then there was the deeper problem—the same nature-nurture argument that had divided the field for decades. DeLisi's program was really run by psychoanalysts, not medical psychiatrists, as she had hoped. During her residency, people like her who were interested in schizophrenia were allowed to take their third year at Chestnut

Lodge. Frieda Fromm-Reichmann's old command post was still in business, a few miles down the road in suburban Maryland, and the therapists there still considered childhood trauma to be one of the main influences in the development of serious mental illness. So did many of DeLisi's teachers at St. Elizabeths.

DeLisi kept reading what she could find about the biology of schizophrenia, and she kept seeing the same name affiliated with NIMH: Richard Wyatt, a neuropsychiatrist who explored not therapy but the effects of mental illness on the brain itself. Wyatt's lab was across town from the rest of the NIMH psychiatry program in the William A. White Building, a century-old red-brick structure on the St. Elizabeths campus with enough room to house patients for long-term study. In 1977, toward the end of her residency at St. Elizabeths, DeLisi went to see him about a fellowship. Wyatt was less than encouraging. He'd see what he could do, he told her, but he usually pulled his fellows straight from Harvard. What's more, he said, no one was going to believe that a mother of two could handle the demands of the job.

This time, DeLisi wasn't angry, just dejected. Even though she had arranged to extend the length of her residency, she had worked twice as hard so that she could finish on time anyway. She was every bit as good as any man, from Harvard or anywhere else. How many of the men in Wyatt's lab had children? Did anyone ever ask that?

DeLisi's mentors in her residency program couldn't understand why she was so depressed. If she really wanted to study schizophrenia, they said, why not spend the final year of her residency at Chestnut Lodge?

Then came the surprise that would launch her career. Wyatt came back with an offer. If she wanted, she could complete her residency with a year in his lab. He'd be getting an extra employee for free, so this was hardly an imposition. If she did well enough, he said, she could continue there the following year as a fellow. No promises, but she could apply.

"I can get you in the back door, if you want," Wyatt said.

ON THREE FLOORS of the William A. White Building, Wyatt had separate labs for brain biochemistry, neuropathology, and electrophysiology; a sleep lab, and a lab where some of his staff collected postmortem human brains for study; and even an animal lab where others experimented with brain tissue transplantation. Wyatt's primary interest was isolating biochemical factors affecting schizophrenia: blood platelet and lymphocyte markers and plasma proteins that might trigger psychosis or delusions. He had research wards with human subjects, each ward holding between ten and twelve patients, referred from all over the country to try experimental medications. Those patients were attended to by research fellows like DeLisi.

Most of Wyatt's researchers were taking advantage of new CT scan technology to search for abnormalities in the brains of schizophrenia patients. These researchers already were finding enough physical evidence of schizophrenia in the human brain that many were ready to turn their back on the environment altogether as a cause of or contributor to the disease. In 1979, Wyatt's team published research showing that people with schizophrenia had more cerebrospinal fluid in their brain ventricles—the network of gaps in the tissue of the brain's limbic system, where the amygdala and hippocampus are located. This was the part of the brain responsible for, among other things, maintaining a sense of awareness of your surroundings. The larger the ventricle size, the more resistant the patients seemed to be to neuroleptic drugs like Thorazine. Here was yet more proof that the illness was physical, not environmental. Or as one of the ventricle study coauthors, a psychiatrist at St. Elizabeths named E. Fuller Torrey, put it: "If bad parenting caused any of these diseases, we'd all be in big, big trouble."

The only problem was that there was no way of telling whether enlarged ventricles were a cause or an effect—something patients were born with, or a condition they developed after they had the illness, maybe even as a side effect of their medication. That, DeLisi had thought, was where genetics would prove to be crucial. But the issue with studying the genetics of schizophrenia in 1979 was that most researchers considered such a thing to be little more than a fishing expedition. Schizophrenia was like Alzheimer's or cancer—clearly the product of more than one gene, perhaps dozens working together—and therefore far too complex for genetic analysis, given how rudimentary the technology for such a search was at the time. This was why the Wyatt lab was focused on more available methods—MRIs, CT scans, and most recently PET scans. So for a time, DeLisi worked on that, too.

Her time with Wyatt was not without tension, and even conflict. DeLisi remembered feeling intense pressure to produce results that would lead to a prize-worthy study. More than once, she felt exploited, like the time she was asked to take the blame for a male colleague's mistake because he was up for promotion. She said no; she had an aversion to going along with the boys, even when refusing did not serve her well politically with Wyatt. "It took two years before he talked to me like he talked to the men in the lab," DeLisi remembered. "They would barge into his office and talk with him, and I could never get time with him."

Finally, at around the same time that DeLisi successfully disproved a long-held hypothesis of Wyatt's concerning the efficacy of a particular drug in treating schizophrenia, an abashed Wyatt remarked, in full view of the men, that she had surpassed them. "Which was nice of him to say," DeLisi said. "But it didn't make the men happy."

EARLY ON IN her time in Wyatt's lab, DeLisi was approached by one of the grand old men of NIMH. David Rosenthal, still on hand as a research psychologist, had decided that it was time to do a follow-up study of the Genain quadruplets, to be published twenty years after the first one. The sisters were fifty now, all four of them still alive. This time they would be brought back for a battery of new biological tests, to see what else they had in common.

DeLisi was pleased by the chance to study the physical roots of the disease. She enjoyed being with the sisters, watching one of them say something and the next one parrot it a moment later, and then the next one, and the fourth one. She conducted CT scans, EEGs, and blood and urine studies. But for DeLisi, the real impact of spending time with four identical sisters with variations of the same illness was that she became more interested than ever in studying genetics.

The only schizophrenia researcher at NIMH who was looking closely at genetics was Elliot Gershon. In 1978, Gershon had coauthored a paper that outlined the best way to verify a genetic marker for mental illness. His idea was to study families with more than one member with the disease. Gershon called these families "multiplex families." The key, he said, was to focus not just on the sick people in the family, but everyone—ideally more than one generation. If researchers could somehow find a genetic abnormality that appeared in only the sick members of a family and not the well ones, then there it would be: the genetic smoking gun for schizophrenia.

DeLisi went to see Gershon and brought up the Genain sisters. Here was a family, the darlings of the NIMH schizophrenia wing, loaded to the brim with the illness, back in town and ready to be examined all over again. "What kind of studies would you do?" she asked.

Gershon's answer brought her up short. "I don't want to be part of this," he said.

DeLisi asked why. She realized how off base the question was as soon as he answered.

"You've only got an N of one," Gershon said—just one set of data, with no variation. "You're not going to find anything really meaningful."

Since all four sisters had the same genetic code, there would be nothing to compare or contrast. This was why Gershon saw no point in studying them. Families were the place to look, he said—but the right kind of families, with varying mixtures of the same genetic source ingredients. The more afflicted children, the better.

If DeLisi was up for tracking down families like those, Gershon said, maybe that would be something he could get behind.

DON
MIMI

DONALD
JIM
JOHN
MICHAEL
RICHARD
JOE
MARK
MATT
PETER
MARGARET
MARY

CHAPTER 19

One of Mary Galvin's earliest memories—from when she was about five years old, in 1970—was being in her bed late at night, trying to sleep, and hearing her oldest brother, Donald, home from the hospital, wailing in the hallway outside the door to her parents' bedroom.

"I'm so scared," he was saying. "I don't know what's happening."

She remembered her parents trying to talk to him, telling him that everything would be all right—that they would find a doctor and figure out what was wrong.

And she remembered Donald running away from time to time—most often to Oregon, to find Jean—and her parents having to track him down, then send him plane or bus tickets.

And she remembered Donald at night again, this time terrified, yelling for everyone to get to safety. There were people in the house, he was saying, people trying to hurt them all.

She remembered believing him. Why would he lie?

MARY WAS DIFFERENT from her big sister. Margaret was tender, empathic, and emotional; she would witness her family's difficulties and internalize them, hardly able to bear the pain. Mary, meanwhile, may have been every bit as vulnerable, but she was also more prac-

tical, shrewd, and, perhaps by necessity now, independent. In first grade, she had been the only child to raise her hand for George McGovern and not Richard Nixon in her class's mock presidential election. Later on, when she got caught with a cigarette at school, her mother asked what they ought to do about it, and Mary said, "Put up No Smoking signs."

Once Margaret left home, taken away to Denver with Nancy and Sam Gary, Mary ricocheted between fury and silence. Her sister's absence ate away at her. She could not understand why she was left behind. Her parents had tried to explain that she was not old enough for the private school Margaret attended there, but that meant nothing to Mary. It didn't change how breathtakingly sudden it all had seemed to her.

In the fifth grade in 1976, Mary was all but alone—watching the fights among the brothers still at home. Peter was testing everyone around him, cycling in and out of the hospital and clashing with Matt, the last of the hockey boys still living at home. Donald had moved into Margaret's room, next to Mary's; the idea was to distance him from the other boys, who slept downstairs, but that just made him even less avoidable to Mary. When he wasn't sleeping off his medications, Donald was pacing and gesticulating and talking to himself. Mary was embarrassed, snapping at him. When that didn't work, she pleaded. And when that also didn't work, she would cry, but not around anyone else. She spent hours in her bedroom, organizing and reorganizing her closet and her desk drawers, lost in thought, in an attempt to have some sense of control.

Out in the world, as she entered junior high school, Mary was all smiles—popular socially, spending more time at friends' houses than at home. She knew that other children weren't allowed to come to her house anymore. She didn't want to be there, either. So she kept to a routine that would keep her away from Hidden Valley Road for as long as possible—from school to soccer in the afternoons and then evenings and Saturdays at the ballet studio at Colorado College; long visits to the Hefley family, who had a horse with a stall that could always use cleaning; anyplace but home.

Mary's mother, after making herself so vulnerable to Nancy Gary and allowing the Gary family to take Margaret in, had tried to return to her old form, putting on a brave and cheerful face in public. With Mary, Mimi demonstrated the importance of not talking about it, of pretending it wasn't happening—of not crying, not getting mad, not betraying the slightest emotion. The same sort of forced equanimity was expected of all the children. On drives from school to practice, or to the Chinook book shop in Colorado Springs, or to tea with the Crocketts or the Griffiths, Mimi offered Mary no explanation of why the brothers were the way they were, or what they could do to help. The most she would say was that the troubles of an eleven-year-old girl amounted to nothing compared to what her brothers were going through.

When Mary felt most helpless, she found a private place in the Woodmen Valley where she could hide, a few hundred yards from the house, on the other side of the hill in their backyard. The kids sometimes called it the Fairy Rocks. Mary would make believe that the Fairy Rocks were her home—pretending to cook dinner there, go to bed there, and wake up there the next morning, on her own and free.

MARY'S FATHER WOULD take Mary with him to the community swimming pool at the Academy, where he was trying to manage a few laps to help him rehabilitate from his stroke. Don recognized people now, but his short-term memory was still compromised, and he seemed fated never to completely recover it. Where he used to speed-read his way through two or even three books a day, now he watched sports on the television, a device he had once not even wanted in the house. His falconry days were behind him. And returning to work was an impossibility. Sam Gary had thrown him a few consulting gigs in the oil industry, but Don wasn't up to the task.

With the exception of Don's military pension, there was no money coming in. The strain of caring for both Donald and Peter was impossible to deny. But whenever Don tried suggesting that Donald and Peter ought to live elsewhere, Mimi's response would always be the same: "Where would they go?" This was a pointless pantomime by now: Mimi was in charge, they both knew that. But even if what her father said went nowhere, it meant something to Mary that he said it. At least he spoke up, making the case for the well children over the sick ones.

Mary would sit with her father as he watched golf on TV and look at him—his memory often working at half speed, his energy sapped—the only person willing to see her situation clearly, to sympathize, to take her side. Alone with her father, Mary would ask him why he was still a devoted Catholic—why, after everything that had happened, he still believed in God. This was not an abstract question to her; she still had to go to mass every Sunday, and she wanted to know what point there could possibly be to that now.

Don told her there had been many times in his life that he, too, had doubts. It was through his own reading and his own intellect, he said, that he found a way back to God.

He did not encourage Mary to do the same. He knew she was not one to be pushed.

SOMETIMES IT SEEMED to Mary that her family had been cleft in two: not the crazy ones and the sane ones, but those still at home and those who got out. Among those still at home, her brother Matt was Mary's soccer coach, and something of a guardian, her defender. Mary once wrote a school essay about him, anointing him the person she most looked up to. But in the spring of 1976, Matt graduated high school and left home, too. Then it was official: Just Mary and two of the sick ones, Peter and Donald. But Hidden Valley Road remained the primary way station for all the sick boys, their one reliable option when they were not welcome anywhere else—even Jim, when he was on the outs with Kathy.

It was all on Mary's mother to choose the right treatments, to search for a solution, to protect them all. Mimi still deeply believed in a miracle. And, for a time, she thought she'd found one, courtesy of a pharmacologist out of Princeton, New Jersey, named Carl Pfeiffer. Pfeiffer's journey through medicine was unorthodox and, at times, deeply strange. In the 1950s, he was one of a handful of pharmacologists tapped by the CIA to conduct experiments with LSD on consenting prisoners. He went on to chair the pharmacology department at Emory University, but then left traditional academia in 1960 and started to publish a stream of papers, all devoid of standard double-blind testing and all based on the fervent belief that brain chemistry depended on a very particular balance of vitamins to keep a person mentally balanced—combinations of supplements that he was prepared to provide to anyone, for a price.

In 1973, Pfeiffer founded the Brain Bio Center, a private clinic that became his headquarters for several decades. Mimi, who had been reading everything she could get her hands on that suggested ways to improve one's brain chemistry, learned about Pfeiffer just a few years after he'd set up shop. When she contacted him, the pharmacologist was more than eager to travel to Colorado to meet the mother of twelve whose sons were losing their minds. After his visit, he invited the Galvins to New Jersey for a complete work-up.

Everyone who was still at home packed up and traveled east to Princeton. Mary remembered someone checking her nails for white spots and being told she had a zinc deficiency, and her mother hanging on the pharmacologist's every word, taking note of everything. Pfeiffer told everyone who came to the Brain Bio Center that what most people considered mental illness probably could be blamed on nutritional deficits. Even Marilyn Monroe and Judy Garland, Pfeiffer said, would be alive today if they had adjusted their blood nutrients. A psychiatric hospital was just, as he once wrote, a "holding tank." This must have been music to the ears of a mother who felt judged every moment, by doctors—and a husband—who suggested her boys would be better off in institutions.

Back home on Hidden Valley Road, Mimi made a ceramic mug by hand for each child. Each morning without fail, she filled the avocado-colored vessels with orange juice to help wash down Dr. Pfeiffer's pills. On the way to school, Mary would get sick, her stomach on fire from the juice and vitamins. She started sticking the pills in her pocket on her way out the door, and chucking them into the woods as soon as she was out of sight.

IN MARCH 1976—TWO months after Margaret left home—a Colorado state highway patrolman noticed a dark-haired man walking east down Route 24 in the precise middle of the road, toeing the double yellow line and talking to himself as cars veered past him on either side. When the officer asked Donald to step to the side of the road, he refused. When he tried to arrest him, Donald started pushing and shoving. It took several officers plus some local firemen to subdue him. In the Colorado Springs jail, the police learned that he had been off his medication for the last several months.

The police transferred him to Pueblo, where by now Donald was rather well known. The doctors learned that he'd just come back home to Hidden Valley Road in January after some time away. He'd gone to Oregon again to find Jean, only to be told this time that she had joined the Peace Corps. He stayed in Oregon for a while, working on a shrimp boat. When he returned, Don and Mimi agreed to take him in, but only if he went to Pikes Peak Mental Health Center in Colorado Springs regularly for his medication. ("They are also involved with several other male children in the family," according to a report from Pueblo.) Donald agreed, but then refused, becoming what the Pueblo doctors referred to as a management problem. "He and family both agree that he should not be living at home," the report reads, "because of his age and his poor influence on other children in the family."

> He denies that he was having hallucinations but would turn his head frequently and look to the side as if he were listening to a voice. He has many religious preoccupations and talks about symbols constantly going through his mind. One of them he described was that of an infant in which the radiance of God was shining down upon. At several points during the interview, he became very tense and expressed hostile feelings such as wanting to knock my block off. . . .

After a few days, Donald still seemed confused and restless and aggressive—or, as the staff put it, "assaultive, destructive, belligerent, suicidal, hyperactive, over-talkative, [and] grandiose." He was written up for "masturbating openly" and "exposing self," and for wandering into the women's dorms and, once, the women's shower. The doctors at Pueblo calmed Donald with Prolixin, but he still reported faithfully about the symbols and signs flashing through his mind.

Still, he was deemed stable enough to be released back home in April.

———

ON THE WEEKENDS, Jim's son, Jimmy—Mary's nephew, though he was just a few years younger than she was—and Mary formed a little two-member day camp. Jim would tell Don and Mimi that he was taking them to church, and instead they would do something fun—go ice-skating, or to the park. Now more than ever, Mary's Saturdays and Sundays with Jim and Kathy became something her parents counted on. "There would be a crisis," she said, "and Mom would call Jim and Kathy to come and get me."

Kathy became like Mary's surrogate mother. That made Jim, in this scenario, the father.

Jim had been coming to her at night when she visited his house ever since Margaret left, when Mary was about ten. He penetrated her with his fingers and forced her into oral sex, and she tolerated him partly out of denial, and partly out of confusion. She remained passive based on the same calculus her sister had used: because she loved Kathy; because anything was better than being at home; because some part of her grew accustomed to not resisting, to interpreting the acts as affection.

Things changed as Mary entered adolescence. Jim had never stopped hitting Kathy, but now Mary would see it in a way she hadn't when she was younger. There was no way for her to rationalize that as anything other than what it was—ugly and frightening and wrong. But she could not abandon Kathy, and so even then, she kept coming back. She continued to endure Jim for the same reason.

Some part of her understood it had to end. She knew that her body was changing, just like her sister's had. She sensed Jim escalating with her, working his way toward something. She thought about what it might mean if Jim tried to go all the way with her—if that meant she could have a baby.

She tried her best not to think about that. But that information sat there. She could ignore it, but not forever.

DON
MIMI

DONALD
JIM
JOHN
MICHAEL
RICHARD
JOE
MARK
MATT
PETER
MARGARET
MARY

CHAPTER 20

There was a gardener who clipped the hedges, and a lady who did all the laundry, and a German chef to prepare steak and potatoes for dinner. There were seven people on staff, all told, not counting pilots for the plane and instructors for the private ski lessons.

The Gary family lived in Cherry Hills, a secluded neighborhood in the southern reaches of Denver, a world away from the bustling downtown. Outside their home was a full-scale ranch with horses. Their driveway had a Porsche and a Mercedes, and their backyard had a tremendous trampoline. Inside, to the right of the entryway, chlorine and humidity radiated from a turquoise swimming pool with a tornado slide and a bubble roof. The walls of all the hallways were lined with paintings: a Modigliani, a de Kooning, a Chagall, a Picasso. In the playroom was a giant swing and a life-sized dollhouse with bunk beds for sleepovers. Margaret's room had a waterbed. That astounded her until she tried to sleep in it. It took a few nights before she gathered enough courage to ask for a regular bed. They got her one.

Margaret got to know Trudy the housekeeper, a second mother to all the Gary children and their friends, and Katie the laundress, who returned her clothes, clean and folded, to her room every single weekday. And she got to know the Garys' eight children, making friends with Suzy, who was a few years younger than she was and a bit of a troublemaker, and Tina, who was a few years older and a bit of a goody-goody. Margaret accompanied the family on trips to the Florida Keys and Vail, where they had a condo on the main drag and she could walk into any store and buy whatever she needed: ski clothes, new Olin Mark IV skis, lift tickets, even snacks from the candy shop when skiing was over. Nancy Gary never went shopping; the shops came to her. Soon, Margaret was wearing the same Lacoste shirts and rugby jerseys as the other children.

In late summer, the entire family would fly to their house in Montana, a modernist showpiece with one wall made entirely of glass, offering a brilliant view across Flathead Lake to the federally protected Bob Marshall Wilderness Area. On the family's hundred acres, there was a cove with a motorboat for waterskiing and tubing and a Hobie Cat for sailing, a tennis court with a guesthouse where the tennis pro stayed, a cherry orchard open for picking, and a stable for riding. The horses were transported from Denver. The servants came along, too, making all the beds and serving all the meals. In Montana, Nancy Gary functioned as a sort of CEO of children's activities, deputizing Trudy the housekeeper as chief camp counselor, scheduling each kid's tennis lessons and horseback-riding lessons and waterskiing lessons. Sam Gary, still running his oil empire, shuttled back and forth between Montana and Denver on his plane to teach all the kids how to water-ski. He'd sit on the edge of the dock with his feet hanging over, hooking the kids under their armpits with his feet, until the motorboat pulled away, yanking the kids forward.

Margaret's parents would say that they had given her the choice to stay home—not to move in with the Garys. But for Margaret, there had never really been a choice. She was being offered the chance to turn in her resignation as her mother's helper: No more dusting the hutch, vacuuming the stairs, feeding the birds, hauling in groceries, or toasting two loaves of bread for breakfast. She had already said goodbye to her summers dancing in Aspen and Santa Fe; those had ended with her father's stroke and resignation from the Federation. Here was a chance to say goodbye to compulsory attendance at hockey and baseball and soccer games; goodbye to four years at Air Academy High School or, worse yet, St. Mary's; goodbye to gymnastics, where she'd never meshed with the coach; goodbye to track, where there was always someone faster than she was; goodbye to the cheerleading squad, which she would miss least of all.

She was being offered the chance to escape the brothers, Donald and Peter, who might erupt any moment. And the other brother down the road, whom she stayed with regularly, and who came to see her late at night.

It was that last reason—Jim—that clinched her decision. When she was being honest with herself, everything else was just an excuse.

And because that was the reason, going to the Garys never felt like just a good thing to her. No matter how much fun she was having, she could never stop framing what she was going through as some sort of expulsion, or exile. How could it be, she'd wonder, that Jim was still an important, even respected, member of the family, while she was the one who had been sent away?

MARGARET'S FOURTEENTH BIRTHDAY in February 1976 came shortly after Nancy plucked her out of Hidden Valley Road. At home, Margaret typically got something modest—a pair of ice skates, a radio from Spencer Gifts. Here, there was a table covered with watches and Frye boots and a full wardrobe to complement her full tuition at the Kent Denver School, the same exclusive private school their own children attended.

Margaret struggled to catch on at Kent. All the kids had their own cars, their own bank accounts, their own allowances and clothing budgets, their own memories of trips abroad with their families to draw from when learning about world history. While Margaret had been going to mass and helping her mother feed a family of fourteen, everyone at the Kent School seemed to have been learning how to throw pottery and silk-screen T-shirts. They seemed so much more artistic and inventive than she was, so free with their impulses. She got cut from play tryouts, got a C on her creative writing project. Their sculptures looked like Giacomettis. She spent most of her first year torn between gratitude and terror, obsessing about what people thought of her. She told herself that the girls who were ostracizing her were just snobs, even as she compared herself to them.

One of the first books Margaret was assigned at her new school was *Great Expectations*. That turned out to be too on the nose for

Margaret, who, like Pip, found herself the recipient of charity from a mysterious benefactor. In Margaret's case, the mystery only deepened because of how friendly the Garys were, how ready they seemed to share what they had. Her dynamic with her host family confused and unmoored her. Once, on just another typical day in Montana, Nancy was cutting up a chocolate cake and started clowning around by cutting off another slice for Margaret or others or eating one herself, and then another slice, and then another, and another, all under the pretense that she was trying to even the cake out. Margaret laughed. It was funny. But in time, she came to understand that this was not her cake, just a gift, and it never would be anything other than that.

SAM GARY WAS about the same age as Don Galvin. Like Don, he grew up in New York, only instead of the outer reaches of Queens, Sam grew up on Park Avenue. He served in the Coast Guard after the war, and during a night off he met Nancy at a dance in Greenwich, Connecticut. In 1954, just a few years after the Galvins moved to Colorado Springs, the Garys moved to Denver, in the midst of an oil boom.

Like Don, Sam was friendly, likable, and quiet and modest in person. But while Don was professorial, Sam was entrepreneurial, a born salesman. Nancy's primary image of her husband in the 1950s was of him sitting on someone's front porch in a rocking chair, shooting the breeze with the owner of a piece of land where he wanted to drill. "At the end, Sam would say, 'And how about leasing your north forty?' or wherever, and they'd say, 'Sure.' So he was very good. He's good with people."

He was also a natural risk-taker. For years, he had been known around Denver as Dry Hole Sam, an impetuous wildcatter with a knack for drilling in all the wrong places. In the mid-1960s, when everyone in the oil exploration business was drilling in Wyoming, Sam started drilling just north of the state line in the southeastern corner of Montana. Sam drilled thirty-five dry holes. He swore off the whole project more than once, but kept turning around and digging again. In 1967, Sam put together another deal to drill on forty thousand acres of land that everyone else in the business was certain had nothing beneath it. He ended up owning most of that project, he said later, "in large part because I couldn't sell any more." He had a helping hand in gaining the right to drill on this land from his good friend at the Federation, Don Galvin.

Don's main role at the Federation had been to smooth the path between the regulators in Washington, D.C., and the entrepreneurs who wanted to invest in business in the American West. When Don needed support for an arts or culture program, Sam was one of the people he'd turn to. And when Sam needed investors for his latest wildcatting project, Don would help Sam make some of those connections. Most crucially for Sam, Don passed along whatever information he heard in Washington about federal land leases and when they were due to expire. On June 29, 1967, one of the new wells—Sam's thirty-sixth try—struck oil in Bell Creek Field in Montana. Sam set up four hundred new wells, hanging on to 30 percent ownership. Which was how, first slowly and then quickly, Sam became one of the richest men in the Rockies. Neither Don nor Sam ever said explicitly that Don had steered to Sam the fateful land lease that had made him rich. But where they once had been casual friends, they became closer after Sam struck oil.

In time, Margaret learned that there were guarded precincts in the Garys' lives that no one could ever penetrate. They, like the Galvins, were dealing with their own family illness, myotonic dystrophy, an incurable genetic disease that eats away at the body's muscles. Four of Nancy and Sam's eight children started to show some symptoms when they were still young and would eventually die from the disease as young adults. The difference was that despite their troubles, Sam and Nancy seemed determined to live their lives with a guileless sense of adventure, gathering their family and friends for backpacking excursions and ski trips. The money helped: Their new wealth allowed them to wear at least some of their burdens lightly, and they also shared what they had. Margaret was not the only child the Garys had taken in. There was a boy the Garys had met on a trip to Mexico, and another girl from Denver. Sam was open about his philosophy of life—how while he may have worked hard, he also felt lucky, and so he felt the need to help those who needed it when he could.

A fifth child of theirs was treated for a time at the private Menninger Clinic, which specialized in schizophrenia. That was an option that Nancy and Sam must have known Don and Mimi could never afford. But there were limitations to their help, of course. They weren't going to help all the Galvins, and so they took one girl—the one old enough to attend the Kent Denver School.

Even during some of Margaret's most comfortable moments there, her thoughts—her greatest enemy now—turned to the nature of that charity. Her mind started playing what-if games that made her feel more and more like she was walking on eggshells. What if Sam had never asked her dad to help him find those government wildcatting contracts? What if Sam had given up on drilling for oil on the thirty-fifth try, and never made his fortune? What if she had never been taken from her home? And the fact that all this had happened—was it because Sam and Nancy really wanted to, because they really liked her? Or because they felt guilty?

INEVITABLY, SHE ACTED out. She started stealing small things to make up for the fact that she felt like she had nothing compared to everyone else. When she raided Suzy's piggy bank, Trudy caught her, but she was not punished. This became just one more thing for Margaret to feel guilty about, to be indebted to the Garys for, and for the Garys to overlook out of a sense of generosity.

But slowly, she assimilated. After years of field trips and river trips and expeditions into the San Juan Mountains, she became a Telemark skier and an accomplished hiker and backpacker. The Kent School boys ignored her until they saw that she was a good athlete. Becoming one of the guys didn't ingratiate her with the girls, but it was something. Her first boyfriend at Kent was someone popular enough to open doors for her socially. With him, she moved on from pot to opium, the drug of choice at Kent at the time. She tried cocaine at an Eric Clapton concert at Red Rocks. She collapsed after too many hash brownies at a Kenny Loggins show at the University of Denver.

She had sex with that boyfriend, too. After what she'd been through with Jim, this felt like an attempt to feel normal, to feel loved. She spent more energy than she admitted fending off the shame of her family's illnesses, and trying to forget everything that Jim had done to her.

She told none of her Kent friends that one of her brothers had died, or that three others were revolving-door regulars at a mental hospital. For those secrets to remain secrets, Margaret could never explain why she came to live with the Garys. She had a stock line about the educational opportunity that Kent offered her, and how lucky she was to have that chance. Covering up the truth might have made her seem fake to some of her classmates. But it was what she needed to do to get through the day, to build some sort of life she wouldn't feel bad about, to survive.

Hidden Valley Road was both home and not home now. Margaret's family seemed apart from her—which relieved Margaret, even as it provoked spasms of guilt. When her parents came for visiting day, rolling past the Mercedeses in their prehistoric Oldsmobile, Margaret flushed with embarrassment. She saw her mother's clothes differently now. She returned to Hidden Valley Road only on holidays, which tended to be the worst times to visit, with every sick Galvin boy stuck in a house together. One year, Matt had to go to the hospital with a concussion after Joe back-flipped him on the patio. When Matt's head hit the concrete and blood started rushing, that only seemed to wind the brothers up more. With barely a pause, another fight broke out downstairs, this one forcing Don to end it. Don, of all people, who was still recovering from his stroke but too furious not to try to do something to contain the chaos.

Margaret remembered the wooden door to the garage broken into pieces, and the ghostly silence once the fighting finally stopped—only after the ambulance came to take Matt away.

IN 1976, MATT enrolled at Loretto Heights, a local private college in Denver, not far from the Garys', to study fine arts. Don and Mimi's ninth son—one of the hockey boys, four years older than Margaret—was a potter, and a good one. Even Mimi said so. He also received encouragement from Nancy Gary, who had served on the Loretto Heights board. The Garys told Matt he was welcome to drop by sometime.

One day, Matt brought a vase he had made to the Garys' house, to show them what he could do. Margaret heard a commotion downstairs, and up came Matt, completely naked. He had taken off all of his clothes, then taken the vase and smashed it. With some of the others, there at least had been some warning signs. But Matt's breakdown was a stunner. It was as if whatever had been slowly overtaking her brothers was picking up speed now.

This was Margaret's old world crashing in on her new one—a reminder she did not belong there, and that nowhere was safe. It was only a matter of time, she felt, before her Kent School friends learned the truth about her family—about her.

CHAPTER 21

"There is a loud telepathic signal here," the skinny man said, calling out with an easy smile to the swath of tie-dye all around him. "If you just be quiet for a while, you can feel it."

Stephen Gaskin was a six-foot-four ex-Marine with a blond goatee, a receding hairline, and a long, untamed tangle of hair flowing down his shoulders. With his time in the military years behind him, he had moved up in the world, becoming a certain kind of prophet. Gaskin had first drawn a following in San Francisco in the late 1960s with a lecture series called "Monday Night Class," in which he regularly filled a two-thousand-capacity ballroom with discourses about acid trips, supernatural activity, and the right way to pursue peaceful social transformation. In 1970, he decided to take Monday Night Class on the road, and he and some four hundred followers traveled the country in a convoy of sixty buses that won them a raft of nationwide media attention. The sign on the convoy said it all: OUT TO SAVE THE WORLD. After wandering across the continent and back again, Gaskin's new community—a nomadic tribe of mellow revolutionaries—paid nearly $120,000 for 1,700 acres of land in the woods of Summertown, Tennessee, settling there in the spring of

1971. Within a few years, the Farm, as Gaskin named it, became the nation's largest commune.

Michael Galvin first arrived at the entrance of the Farm in 1974, partly as a hippie in search of a new way of living, and partly because he was out of options. The tragedy of Brian and Noni had brought low everyone in the family, but it was Michael who had gone with Don to see the body, and Michael who stood there as a police officer explained in cold, clinical terms what had happened to his brother—and to that poor girl. He still believed that, in an alternate reality, he might have helped his brother—how, if he had gone straight to Sacramento and not detoured in L.A., he might have made it there in time to do something. What, exactly, he couldn't say.

Mimi and Don must have sensed how difficult this was for Michael. They decided to send him east to New York to stay with an uncle, Don's brother George, who worked as a conductor with the Long Island Rail Road. Michael's parents thought that he might find Michael a job as a brakeman. When Michael failed the engineering test, he went to see his maternal grandmother—Mimi's mother, Billy, then living in New Jersey—who came up with another idea.

AT THE HEIGHT of its popularity, the Farm attracted a population of about 1,500 people. Michael might have been the only one who pulled up to the front gate in a Buick driven by his grandmother. Before being let inside, Michael was informed of the rules. No overt anger. No lying. No private money. No eating animal products. No smoking tobacco. No alcohol. No man-made psychedelics like LSD. No sex without commitment (Stephen Gaskin was licensed by the state of Tennessee to perform marriages, and did so frequently, preferring to marry two couples to one another in what he called a "four-marriage"). Michael said yes to it all.

Despite Gaskin's wholehearted endorsement of tantric sex and the bountiful supply of homegrown hallucinogenic mushrooms, Michael learned that the Farm was not a place where anything goes. Behavior was always policed, often by Gaskin himself, who would complain that all he had time for all day was settling everyone else's conflicts. And for a bunch of anti-authoritarians, the Farm's inhabitants had one leader whose rule was never called into question. Gaskin controlled what drugs people took, who slept with whom, and how money worked in the community (whose members relinquished cash, cars, property, even inheritances to the cause). Gaskin became known for meting out banishments called "thirty dayers," during which the Farmies were supposed to get their heads right. "A smart horse runs at the shadow of the whip," he once said. He required some people to take a vow of chastity, even as he had three wives of his own that he shared with two other men—his own "six-marriage." One of those wives, Ina May Gaskin, would revolutionize natural childbirth in America with a book published in 1975, *Spiritual Midwifery*. Four or more babies were born at the Farm every month, keeping Ina May and her trainee midwives busy. "Farmies," she would say, were "a special kind of hippie: they worked."

Michael found that he didn't mind the work. In a weird way, he kind of craved it. Gaskin had always insisted that the Farm was not a cult but a collective—a demonstration project for a different way of living. His lectures touched on the teachings of the Tibetan yogi Milarepa, whose own master cast him into the depths of despair in order to mold his character. The key wasn't to tune out like a stereotypical hippie, but to notice what was happening around you—to hear the signal. "If you get too used to it and don't pay attention to it, it's like living by a waterfall," Gaskin said. "People who live by waterfalls don't hear them."

Gaskin's Sunday morning talks, mass meditation sessions attended by the entire Farm community, were more meaningful to Michael than any Catholic mass he'd ever attended. Michael received validation and confirmation of things he had only suspected—that science only describes the physical world, not matters of the heart. He loved how Gaskin always said "keep closure": If you leave someone hanging, be sure you go back to them and make sure everything is understood between you both. On Hidden Valley Road, there had never been closure, just sibling rivalries layered on top of one another. Even his father's attempts to get everybody to live in harmony never worked. Instead of clearing the air, they'd watch a football game. Could it be there was another way to live?

The most intense moments for Michael took place in a tent called the Rock Tumbler. Set off away from the community, the Tumbler was where men whom Gaskin considered too oppositional were sent to dissect one another's troubles—*We need to talk. What are you doing?*

Why are you doing that?—until, so the theory went, their rough edges were smoothed out. Gaskin doled out "constructive feedback" for Farm members who were "on a trip"—too uptight or angry, or not empathetic, or too lazy. "You are the only variable in the situation you have control over," he would say. "If you're not grooving all the time, find out why you're not grooving and fix it."

Michael had never experienced anything like this before. Everything in his own family had been so top-down, so dictatorial, with a pecking order that invited the older siblings to victimize the younger ones. Here there was a leader, sure, but the community acted on consensus to hold everyone accountable, and to dig and dig until the subconscious issues at the root of the problem became known to everyone.

This was a Watergate-style inquiry: The denial and suppression and cover-up of a problem were as bad as the problem itself.

Michael ended up loving the Tumbler. Everything about the Farm felt wholesome to him—progressive, well-intentioned people being good to one another. But while he was there, his contempt for his own family only intensified, sometimes even overpowering everything positive he was feeling in the moment. He hadn't gotten over how his parents once wanted him committed. He knew he wasn't insane; what system, what family, would send him to the hospital that way? Don and Mimi were so repressive, he was convinced that they were part of the problem.

At the end of eight months, Michael and his Tumbler-mates had become isolated—so much so that Gaskin commanded them to disband their tent and come back to live closer to the heart of things. Michael went under the tent to grab a bag he'd left there, and when he opened the bag at his new quarters he saw thousands of tiny bug eggs, spilling out of the opening.

Michael took that as a sign that his time at the Farm might be winding down. He went to Gaskin and said he needed to get away. There happened to be a bus leaving for Albuquerque, so he took it, taking a new set of tools for living with him.

HE WAS NOT ready to go home just yet. An old friend was heading to Hawaii. Michael tagged along, finding a $130 air fare from L.A. He stayed for about a year, finding short-term work hanging drywall, living off of food stamps, being out on his own completely, without his family or his surrogate Farm family.

He moved a little bit further through his grief. And he was about to move on to the Philippines with a new friend when his mother, on the phone, told him that she missed him and that she wanted to send him a plane ticket.

Here was the chance to put the Farm's lessons into action for the family he'd left behind. Michael came home to Colorado Springs and enrolled in a community college to learn mechanical drafting. But he had returned to even more conflict than he would have anticipated. Donald was there, and Michael found himself infuriated by him— why wasn't he making choices that were helpful to him? Was he too far gone to be saved? Things were even worse now than before he left. Peter was sick, too. Their father had his stroke. Everything seemed more out of control than he'd remembered it. And no one was taking any of his advice. He wanted them all to eat brown rice and meditate, and they wouldn't have any part of it.

Michael came away dejected. What would it take for his brothers to do what he'd managed to do? When would they learn to get out of their own way? When would they notice the waterfall crashing around them?

DON
MIMI

DONALD
JIM
JOHN
MICHAEL
RICHARD
JOE
MARK
MATT
PETER
MARGARET
MARY

CHAPTER 22

Mary never stopped campaigning to visit Margaret. Her parents let her spend weekends in Denver every few months, when the Garys hadn't jetted to one of their other homes. When summer came, the Garys also paid for Mary to spend two weeks at Geneva Glen, a sleep-away camp that ran its campers through a number of elaborate imaginary scenarios—the Knights of the Round Table, Native American traditions. For the first time in her life, Mary, away from her family and away from Jim, had permission to let down her guard, remove her mask a bit, and forget about what was going on at home. At the end of her first session in the summer of 1976, Mary called home, begging to stay. The Garys paid for her to stay the full eight weeks. She went back every summer until college.

For a couple of weeks at the end of every summer, the Garys opened up their Montana place to a platoon of kids, including friends and cousins. Mary was there for that. She and Suzy Gary were mischievous kindred spirits, sneaking sips of Sam's Coors Lights. It still confused Mary that Margaret got to live in this world all the time while she had to beg and plead for the chance to visit. But as Mary got older and mixed with the Garys a little more, she and Sam started to have longer conversations about her future. Whenever Mary said she wanted to do something for the greater good, Sam's response was always the same: "If you want to do something like that, go make some money and give it away."

Both Mary and Margaret loved the Montana trips. But while for Margaret, Montana was another place where she never truly felt at home, for Mary it was a taste of what life could be like if she didn't have to be at home at all.

MATT CONTROLLED THE stoplights in Colorado Springs for a long while. Then he announced that he was Paul McCartney.

After his breakdown at the Garys' house, Matt had dropped out of his ceramics program at Loretto Heights in 1977 and was back home now with Donald and Peter. Mary—twelve years old and the only sane child living at home—no longer had Matt as a protector. Now he was part of the problem, a hazard. One day, Peter was being a pest to Mary, and Mary had asked Matt for help. Her parents were not there; neither was Donald. The two brothers faced off in the living room, the same way Donald and Jim used to. Once the punches flew, the pretext for the fight didn't matter. Both Matt and Peter lost control, each of them accessing something primal, something Mary hadn't seen before. She was sure they were going to kill each other.

There was only one established move in these situations, a move that Mary knew well by now. She rushed to Don and Mimi's bedroom, flipped the lock behind her, and called the police. That was when Matt turned on her; the last thing he wanted was the police at their door. She sat there, trembling, the phone in her hand, as Matt, once the brother she admired most, tried to break the door down.

The police arrived before Matt could get to her. They took Matt away to the hospital. For Mary, this was the first time she'd felt responsible for hospitalizing one of her brothers. She was surprised, after so many years feeling rage toward them, to feel guilty about that.

She was also surprised that she had actually not wanted them to hurt one another—that after building up so much resentment toward them, she still cared.

Matt's first admission to Pueblo was on December 7, 1978. Five days later, Peter joined him there, for his third visit to Pueblo that year. Donald also was cycling in and out of Pueblo that year—three Galvin brothers on separate wards of the same hospital, for what would not be the last time.

From then on, when Mary was alone with Matt and Peter, she locked herself in her parents' room until someone else came home.

PETER WAS THE closest brother in age to Mary, just four years older. At home now, Peter was a wall of *no*—he refused all help and defied all advice. He never thought he needed medical care. It followed that he did not believe he needed his shot of Prolixin every three weeks.

By 1978, the year Peter turned eighteen, the staff at Pikes Peak knew the whole family well, especially Mimi, who had become a fierce advocate for each of the sons. Between outpatient visits, Peter would stay at Hidden Valley Road only as long as he could stand it, or until he became too much for his parents to handle and they sent him away. Then he would camp out under a bridge for days at a time, or hitchhike to Vail and hang out along the main strip.

Peter was in and out of hospitals a half dozen times that year. A supportive residence called CARES House in Colorado Springs took him in briefly, but when Peter left without permission, the staff said he was not welcome back. On July 2, an argument with his parents over taking his Prolixin ended with Peter smashing four picture windows. Peter later explained that he "really did not want to get into a hassle, but it just happened." Once again, his parents threw him out of the house; this time, he was old enough to be sent to the state mental hospital at Pueblo.

Over three stays at Pueblo, the staff got to see both sides of Peter. He could be charming—"a pleasant, alert, oriented and well-groomed young man who behaved appropriately in the interview situation." But once the conversation turned toward his family, "his overall style was markedly grandiose and paranoid" and later "belligerent" and "very hostile." Peter announced that he had an interview for a job at the Eisenhower Tunnel; then he said he had decided to start work as a ski instructor in a few weeks; then he mentioned that he'd recently done some work as a stunt skier for the TV show *Charlie's Angels*. At times, the staff at Pueblo needed to put him in restraints; then, once the restraints were removed, he would decide to leave the hospital.

Once he made it as far as Ordway, a tiny town of a thousand people fifty miles east of Pueblo, where he jumped on one car and tried to leap onto a moving truck and was almost run over. Another time, he said he was a Secret Service agent, working for the Queen of England. "Presently, Peter is so loose and psychotic," one report read, "that interviewing him is fruitless and nonproductive."

For perhaps the first time, the doctors, struck by "his irritability, his demandingness, his mild hyperactivity, [and] his manipulativeness," suspected that Peter's problem was likely not schizophrenia at all but bipolar disorder. If that were the case, that revised diagnosis would cause an entirely new set of problems: Peter was too unreliable to be trusted to regularly take lithium, the drug most prescribed at the time for that condition. Lithium is one of the few psychiatric medications that is dangerous in mild overdose; Peter would not only have to follow the drug regimen, he would have to agree to have his blood level monitored, and that didn't seem likely. As long as he stayed on Prolixin, he seemed more or less all right. So they decided to stick with schizophrenia as a diagnosis, concluding that "the distinction does probably not have any practical importance at this time."

For the next several years, Peter would be prescribed drugs to treat schizophrenia, when it was quite possible he was suffering from another illness altogether.

WHEN DONALD WASN'T at home or getting outpatient treatment at Pikes Peak Mental Health Center in Colorado Springs, he was still walking upward of two hundred miles a week. Jobs would come and go, but walking remained his great constant, along with religious visions and preaching. Only now and then would his wanderings get him into trouble. He was brought back to Pueblo in September 1978, after a squabble with a clerk at a sporting goods store. During that nearly three-month stay, he announced plans to leave the country at Christmas and give up his citizenship.

He returned to Pueblo a year later after having an argument with a nurse at Pikes Peak. That was when he started talking about various stars in the sky showing him where to find particular elements in the ground that were involved with what he called "rock knife chemistry." He believed that he had to find those elements, smash them with a hammer, and eat the dust.

Donald was discharged on January 7, 1980, only to be readmitted in March—his sixth visit to the state hospital in Pueblo in ten years—after Don and Mimi lost patience with him and told him to get his own apartment. On the ward, Donald shouted about Jesus, and his Thorazine prescription was increased several times with little effect. He was discharged in June, once he was stabilized with an antipsychotic drug called Loxitane.

But he returned yet again in November. He stopped taking the medication and had been staying awake for eighteen hours a day, walking around naked in the house, screaming at the top of his lungs. Jean was back in his thoughts. He referred to her as his wife. He also was talking about guns and knives.

Mimi and Don, according to the hospital report, were afraid of their oldest son. "They want Donald to get the strong message that they love him," the report reads, "but they cannot accept him until he has been stabilized on the medication."

JOHN, THE MUSIC TEACHER, was in Idaho. Richard, once the schemer of the family, was trying to start a business in Denver. The two hockey brothers who had not become sick, Mark and Joe, also were hours away in Boulder and Denver—there but not there, able to avoid the worst of it. Mark, once the family's prodigious chess master, was hurt deeply by what had happened to his hockey teammates, Matt and Peter. Joe, driving a fuel truck at the airport, was living quietly, even as he seemed to be exhibiting some of the warning signs of psychosis—a disconnection from everyday life, problems understanding basic social cues.

And then there was Jim.

The most important rule of the house was clear enough: The last thing Mary should ever do was talk about any of this. But she saw what was happening to her family. She was angry about that, even as some small part of her was preparing to be next. As she got older, Mary stopped hiding her frustration. She was almost thirteen—not a little girl anymore, and not to be trifled with. She banged on her walls at night now with abandon, without apology, trying to get Donald to be quiet.

She noticed other changes, too. During the day, she sensed a growing distance between her mother and father. It was as if Mimi had become her husband's caretaker now, nothing more. Once, her mother even left for a few weeks, staying with her sister, Betty, back east, leaving Mary alone with her father and brothers. Another desertion, another abandonment.

Mimi must have noticed that Mary felt this way—recognized the anger inside her, maybe even identified with it—and started taking her on shopping trips downtown, just the two of them, and to tea parties with her friends. Without explicitly saying so, Mimi was working to ingratiate herself with Mary—to let her know she loved her, too. Despite herself, Mary found herself enjoying this time with her mother, away from the others. While she thought all she wanted was to get away, what she really wanted, perhaps, was this sense of closeness—an uncomplicated love, free of mystery, free of danger.

DON
MIMI

DONALD
JIM
JOHN
MICHAEL
RICHARD
JOE
MARK
MATT
PETER
MARGARET
MARY **LINDSAY**

CHAPTER 23

MARY HAD TRIED to follow Margaret to the Kent School. When her seventh grade application was denied, she was furious. *I can't get in to Kent?? My* sister *is at Kent!*

At the start of eighth grade, in 1978, Mary told her father that she wanted to go to boarding school. Don asked Sam Gary for advice. Sam asked Mary if a place like his alma mater, Hotchkiss, in Connecticut, might appeal to her. Technically, Sam had been kicked out of Hotchkiss, but all was forgiven now.

Mary didn't hesitate. She'd already been doing whatever she could to stay out of the house. If she got into a celebrated, unimpeachably refined school two thousand miles away, there was a chance she might never have to come home again.

Mary applied to Andover, Exeter, Hotchkiss, and Taft. She got into them all. She chose Hotchkiss because it seemed like the prettiest, the one farthest away from a city. The Berkshires seemed like a reasonable substitute for the mountains of Colorado—the best that she could do.

Mary's tuition was paid by a scholarship funded by another alumnus. The Garys picked up other costs, like transportation. After three years of looking for a way out, Mary had earned her ticket.

THIS NIGHT WOULD be different. Mary knew it had to be.

She was thirteen years old. Jim was thirty-one, still married to Kathy, and still running the Manitou Incline. Behind the funicular station at the top of the hill was a musty cottage with a couple of old mattresses and sleeping bags. As the manager of the incline, Jim had unfettered use of this cottage. Sometimes, instead of hosting his younger brothers and sisters at home, he invited them there, at the top of the incline, where they could be alone.

This time, on a cool evening in the spring of 1979, Mary was there with Matt. Jim had invited them both to camp out and smoke pot and drink beer. When it got late, she fell asleep in one room of the cottage, the guys in the other. Matt was passed out, but the light was still on, so Mary pretended that she was asleep, just as she always did when she knew Jim would come to her—disassociating by pretending it wasn't happening, at least not to her.

But she could not go through with it that night. Mary had gotten her period. She was more terrified of getting pregnant than she was of Jim's fury at being refused.

For the first time, when Jim came over to her, she lost control, saying things she hadn't expected to say. *Leave me alone. Get away from me. I hate you.*

Jim attacked her anyway. He entered her, something he'd never accomplished with Mary's sister. He came. And he never spoke to her about it after that. He avoided her altogether.

There were, of course, several weeks of terror that she might become pregnant. Once it became clear she wasn't, Mary expected to feel relief. She'd done it: She'd fought him off, protected herself, made it so that he would never do it again. She was almost delirious with the thought of it.

But then, quite unexpectedly, part of her found Jim's ability to disappear from her life to be utterly wrenching. She tried to ignore that feeling, but there was no mistaking it. She was heartbroken. Some part of her had truly believed, as a child does, that this was love.

————

SHE WAS ALMOST free now. Jim was no longer in her life. Soon neither would Peter or Matt or Donald. Her future was her own. At the end of eighth grade, not long after she was accepted to Hotchkiss, Mary was invited to a high school party hosted by the older brother of a friend. She said yes right away.

Mary told her mother she was sleeping over at her friend's house. She left out the part about the party. When she got there, the big brother was there, along with two other guys, drinking Seven and Sevens. She joined in.

The guys invited both girls out to a well-known make-out spot in town to drink some more. Her friend said no; she had to stay home to take care of her little sisters. But Mary said yes and got into a car with them. By the time they came back, Mary's friend and her sisters were all asleep. Mary was so drunk she could barely stagger back inside.

The boys, seeking privacy, found a walk-in closet, opened the door, and directed Mary inside it. One at a time, they followed.

Mary woke up a few hours later with no idea of where she was. She opened the closet door and found her way to the living room. Daylight streamed through the windows. Mary shuddered. Her mother was supposed to pick her up. She stumbled outside and waited on the curb, holding her stomach, trying to sort out what had happened.

The plan had been for her mother to take her to a dentist appointment. "I can't go," Mary said, as soon as she got in the car. "I'm sick." Mimi might have gathered that her daughter had been drinking—this was her twelfth teenager, after all—but she said nothing.

It was then, on the way back home, it all came flooding back—two boys taking turns, a third halfheartedly trying to stop them. Mary almost threw up all over herself. A fitting punishment, she thought at the time, for a girl who had been so bad. She had lied to her mother, and she had gotten drunk, and she had failed to run away.

Too clouded by shame to place the blame on anyone but herself, Mary told no one what happened. She figured everyone she knew would know sooner or later. She made herself a promise that day: Once she left for Hotchkiss, she would never live in Colorado Springs again.

No more teenage boys in closets.

No more Jim in the cottage, at the top of the Manitou Incline.

No more Donald or Matt or Peter or anyone but her and her alone.

—⁓—

It was still orientation. Too soon to be pigeonholed, she hoped. All she wanted was to be the last person anyone at Hotchkiss would ever think was unusual. Then a teacher she'd just met read her name tag and scowled.

"There is already a Mary Galvin at this school," she said. "What's your middle name?"

Mary did not answer right away. She knew that her name said more about her than she wanted anyone to know. Forget, for the moment, how being called Mary Christine had helped to make her, in her brother Donald's eyes, the sacred virgin mother of Christ. Sitting there with all of those sons and daughters of privilege, feeling the East Coast WASP-iness of the place, Mary sensed that her Catholic name screamed *not one of us*.

In a flash, she thought of another name. Thomas Lindsey Blayney was her great-grandfather on her mother's side. Lindsey was a scholar and an eminence for the family—the kindly and wise Don Galvin of his generation. Lindsey had remained in close touch, writing Don and Mimi and doting on his great-grandchildren.

Lindsey seemed like a prep school name to Mary—a better name, a Hotchkiss name. She flubbed the spelling, a mistake that had the virtue of making the name all hers. She had to do something, to make some sort of gesture that would wipe away everything that had happened to her in the first thirteen years of her life.

"Lindsay," Mary said.

And from that moment on, Lindsay was her name.

CHAPTER 24

1979
University of Colorado Medical Center,
Denver, Colorado

Robert Freedman and Lynn DeLisi never worked in the same lab or even the same research institution or hospital. They were just two of hundreds of researchers around the world who were investigating schizophrenia. Their specialties were different, too—two disparate approaches to the same problem. While DeLisi wanted to track down the genetic components of schizophrenia, Freedman was on the hunt for a physiological understanding of the illness. She wanted to learn where it came from; he wanted to learn how it worked.

Neither of them knew that their paths one day would merge in the study of one extraordinary family—and that what they would learn from that family would help them both unearth new knowledge about the disease.

While DeLisi's path to a career in medicine was riddled with detours, Freedman's had been more or less seamless. He graduated from Harvard in 1968, two years after DeLisi graduated from the University of Wisconsin, and entered Harvard Medical School right away. As an undergraduate, Freedman had been drawn to the idea that the human mind could synthesize its own, entirely separate reality. "It just seemed to me if there was ever a disease that was uniquely human and philosophical, it was having schizophrenia," he said. At the same time, Freedman was fascinated by the physical body, particularly the workings of the central nervous system. After medical school, he directed his career toward the study of the brain, starting off with the belief that there must be a better way to learn why neuroleptic drugs like Thorazine did what they did.

Freedman understood from a new flurry of research that people with schizophrenia might have difficulty processing all the information sensed by the central nervous system in an efficient way. This "vulnerability hypothesis"—an update, or elaboration, of Irving Gottesman's 1967 diathesis-stress hypothesis, introduced by a team of researchers from Harvard and Columbia in 1977—sought a middle ground between nature and nurture by suggesting that certain genetic traits directly compromised the brain's sensory and information-processing functions, making the brain especially vulnerable to any number of environmental triggers. To these researchers, those triggers—anything from everyday heartbreak, to chronic poverty, to traumatic child abuse—didn't cause schizophrenia as much as provide "an opportunity for vulnerability to germinate into disorder." And that vulnerability, many thought, was really an issue with "sensory gating," or the brain's ability (or inability) to correctly process incoming information. A sensory gating disorder was the most common explanation for the schizophrenia experienced by John Nash—the Nobel Laureate mathematician depicted in *A Beautiful Mind*—who was able to detect patterns no one else could, and yet also was prone to delusions and visions of beings who were out to get him. Both of those aspects of Nash's personality were said to be products of the same hypersensitivity.

Neurons talk to one another through brain synapses, the junctions between nerve cells that are essential for sending messages through the central nervous system. Many researchers came to suspect that the John Nashes of the world weren't able to prune their synapses in the same way as most people.* Some people with schizo-

---

* In 1982, Irwin Feinberg of the University of California at Davis codified this idea as the "pruning hypothesis." Schizophrenia, he proposed, often first ap-

phrenia, they thought, might become sensitive to distracting sounds and feel flooded by too much information—the way it sometimes seemed Peter Galvin felt, or Daniel Paul Schreber had back in 1894. Others might become hyper-reactive, guarded, even paranoid—like Donald Galvin, mysteriously inspired to move all the furniture out of the house on Hidden Valley Road. Still others might be unable to pick and choose what to focus on with any reliability and might become delusional—seeing hallucinations and hearing voices, like Jim Galvin.

Sensory gating was just a theory. But once Freedman came to the subject, in 1978, as a researcher at the University of Colorado Medical Center in Denver, he started to develop a deceptively simple method for measuring sensory gating—and, by extension, indirectly measuring the vulnerability of a brain to schizophrenia. Freedman realized that the other researchers who were studying sensory gating—measuring their test subjects' reactions to various lights and sounds and such—were skipping right past an important part of the process. As a neurophysiologist, Freedman understood physical reflexes and their peculiar, even counterintuitive relationship with the brain. He knew there were neurons—brain cells—that ordered you to move your muscles, but also neurons that inhibited the movement of those same muscles. In order to walk, for example, your central nervous system needs both kinds of neurons, for action and inhibition. Otherwise, everyone would be falling down all over the place. Why wouldn't it be the same, Freedman wondered, for thinking?

What if the problem with schizophrenia patients wasn't that they lacked the ability to respond to so much stimuli, but that they lacked the ability *not* to? What if their brains weren't overloaded, but lacked inhibition—forced to reckon with everything that was coming their way, every second of every day?

In 1979, working at his lab in Denver, a little more than an hour's drive from the Galvin family's home on Hidden Valley Road, Freedman developed a method of measuring inhibition that was painless for the patients: A small electrode was placed on the test subject's

---

pears during or just after late adolescence because of "a defect in the [brain] maturational process" in which "too many, too few, or the wrong synapses are eliminated."

scalp, and that electrode measured electrical activity in the form of waves. Bigger waves meant the brain was working harder to process information; smaller waves meant the brain was doing less. Freedman devised an experiment. He measured his test subjects' reactions when they heard the same exact noise—a click—played twice, with just a short interval between them, usually half a second.

Any so-called "normal" brain, a brain without schizophrenia, recorded a large brain wave reacting to the first click, followed by a smaller wave reacting to the second click. The normal brain learns from what it perceives. It doesn't have to start from zero if it hears the same thing twice. People with schizophrenia, however, couldn't manage that. In test after test, conducted at Freedman's lab in Denver, their brains showed two waves of equal size for the two clicks. It was as if they had to react all over again to the second click—even though they had just heard the same click a fraction of a second earlier.

The double-click test was not testing for schizophrenia itself. It was testing sensory gating, which was one potential aspect of schizophrenia. What made this result so exciting was that a sensory gating deficiency might well be genetic—and therefore could be traced through generations. Freedman felt as if he were on the cusp of a major breakthrough not just in understanding schizophrenia, but in treating it: What if he could isolate the gene irregularity that caused people to react this way to the double-click test? If he could do that, and if those people were indeed diagnosed with schizophrenia, then he would have proven the existence of a gene related to the illness and opened the door to a genetic remedy.

No one had ever done such a thing, though many dreamed of doing it. This was a common enough strategy for other diseases: With diabetes, for instance, there may be ten or twenty different genes in play, but the first generation of medicine treating diabetes targeted just one of those genes.

All it would take, Freedman thought, was the identification of one gene. What might help him in that search, he thought, was a large group—a family—with an extraordinarily large incidence of schizophrenia.

Where Freedman would find such a family, he had no idea. But they were out there somewhere. Probably closer than he thought.

DON
MIMI

DONALD
JIM
JOHN
MICHAEL
RICHARD
JOE
MARK
MATT
PETER
MARGARET
LINDSAY

CHAPTER 25

The Galvin sisters were both beautiful, with long brown hair and bright eyes and dimples and high cheekbones. When they entered their twenties, they would even model a little, for print ads and outdoorsy magazines; Lindsay posed on skis, up on a mountain ridge, her hair flowing over a purple parka. They had boyfriends, plenty of them. And drugs—pot mostly—but neither of them seemed to take much pleasure in them. Drugs were more helpful for covering up the past and trying to replace it with something else.

As little girls, the sisters had never quite connected. Margaret, before leaving home, was too busy searching for somewhere else to be to entertain a sister three years younger than she was. Lindsay, shattered by Margaret's departure, became jealous of her older sister, angry that Margaret got to leave and she did not. But all that changed as soon as both sisters found themselves on a similar course, away from Hidden Valley Road. *I love that little girl so much,* Margaret wrote in her diary at college, *and she* must *know it—we have a great sisterly relationship—it's so unbelievable that we're so tight.*

Lindsay, in turn, wrote Margaret a poem about the connection they shared now.

She is not there to pass each day
She has become a part of me
She has built, open, found me
Looked within me, found me
Become a part of me
She climbs mountains
I succeed
She inhales the air, I exhale
Nature fills her heart up
And overflows into mine
She is a part of mountains, air and trees and plants
She is part of me
Oh us
She cries as I laugh and laughs as I cry
Her joy, my sorrow
My sorrow her joy
I feel her pain her pleasure feels for me
To be two as one in two different places together
Oh us

Many of her family members were slow in coming around to calling her by her new name. Some, like her mother, never would. But

Lindsay, left, and
Margaret

that was fine with Lindsay. The new name wasn't for them. It was for her new life. But even behind her new guise—adopting a persona, or trying to—Lindsay stood out at Hotchkiss from the start. She had shown up in ninth grade at a school where most students started in tenth grade, and that was enough to get kids talking. Anyone arriving out of sequence had to be going through something strange. Had she been expelled from another school? Were her parents divorcing? Or was there some other drama they could only guess at?

Lindsay stood out in other ways, too. She dressed like a prep school girl, in plaid skirts and collared shirts. She hadn't known that the girls at Hotchkiss were doing the Deadhead hippie thing. And she had grown up with her father's liberal politics, and now she was hearing some of her classmates talking about how anyone on welfare was just riding on someone else's coattails. She found a few sympathetic adults, an English teacher and a philosophy teacher, who didn't mind her barreling into their offices and bursting into tears, crying, *How could they think this?* And she crafted a survival strategy. Obviously, she wasn't going to be going on shopping trips in Manhattan with anyone. She wasn't going to Paris on spring break. Instead, she became an athlete—soccer, mainly, and lacrosse—and that became enough for her to make it through her time there.

Lindsay had been practicing masking her emotions for so long that doing so came naturally to her. Performing in this way—a permanent smile, and an air of personal secrecy—took a small toll. She wasn't getting the straight As she'd expected. But like all Hotchkiss students, she read *Walden,* and Thoreau's transcendentalism was a tonic to her, reaffirming her need to be out in nature—like, of all people, her mother. That she was finally so far away from Mimi only to realize how much she shared with her was, to say the least, a surprise.

Some part of Lindsay didn't think she deserved to enjoy Hotchkiss—that she could pretend to be carefree, but really that state of mind would always be out of reach, reserved for others. Now and then, she would be reminded of exactly how different she was. When she and a friend went to a screening on campus of *One Flew Over the Cuckoo's Nest,* Lindsay didn't last ten minutes. She ran out of the auditorium in tears. Her friend was concerned. When Lindsay muttered something about how there was mental illness in her family, the friend did not ask any more questions.

———

LINDSAY WAS AT Hotchkiss in 1982 when Joe—the oldest of the four hockey boys, the mild, thoughtful seventh son, nine years older than she was—had his psychotic break.

The doctors who had met him when he'd visited Peter a decade earlier had an inkling that something was wrong. But Joe had seemed all right to the rest of the family, or at least well enough to live on his own and work. After high school, Joe had found work at the airport in Denver, and from time to time he would take her skiing, get her out of the house, help her feel normal for a while. Then he got a job with United in Chicago, working as a baggage handler, and he moved there and fell in love with a doctor's daughter. A wedding seemed imminent until Joe was refused a promotion at work. For Joe, this seemed to be the culmination of many insults he'd endured while working there, including a knee injury he'd been nursing that he'd never filed a claim for. He started to send threatening letters to his bosses. When United fired him, Joe sent more threatening letters, this time to the White House.

In short order, Joe lost everything—his car, his apartment, his fiancée. Then he started seeing things. First Donald and Jim, then Brian and Peter, now Matt and Joe—six of the twelve of them, lost.

Lindsay was brought low all over again. She flew to Chicago to join her parents, who were coming to see Joe at a hospital. What she saw horrified her. Joe was drugged, hardly responsive. It dawned on her that she had never visited any of her brothers at Pueblo—never before seen what happened to Peter and Donald and Matt when they weren't at home. For the first time, she started to think not just about their behavior, but the kind of medical treatment available to them.

Joe returned to Colorado Springs, joining Peter and Donald and their parents in the house on Hidden Valley Road. He was hearing voices all the time now. One night, he went running down the middle of a street downtown, screaming at the top of his lungs, "The wolves are chasing me!" It took two six-foot troopers to subdue him. He spent much of May 1982 at the state hospital in Pueblo.

Michael, the hippie alumnus of the Farm, was living nearby now, and was as shocked as everyone else by how quickly Joe had changed. He still suspected that if his brothers had a less repressive upbringing, they never would have snapped. He decided that Joe might not be so far gone yet—and that maybe he could help bring him back. He went home to see Joe and spent a night out driving with him, trying to get him to let out whatever anxieties he had, trying to reach some part of him he was keeping hidden. *We need to talk. What are you doing? Why are you doing that?* He took Joe to a field on the grounds of the Air Force Academy. *Hey, let it out!* Michael remembered saying, over and over again.

Nothing worked. His brother was unresponsive, confused, and often just mentally elsewhere. Michael thought that this must be what it was like to talk to an alcoholic—someone too tied to his current state to imagine any other way of being. He couldn't stop thinking that mental illness was a choice, and that Joe was making the wrong choice.

If Michael was frustrated, Lindsay, back at boarding school, was surprised to find her resentment easing, her rage subsiding. Like Margaret, she had felt marginalized at her exclusive private school—but Lindsay stopped thinking that the solution ought to be to deny her family's existence. Instead, she discovered a certain kinship with her sick brothers. They were ostracized by society. Sometimes she felt that way, too.

—⁂—

Margaret had traveled east in the fall of 1980 to start her freshman year at Skidmore College in upstate New York, a two-hour drive from Lindsay at Hotchkiss. At Skidmore, Margaret experienced some of the same culture shock she'd gone through at Kent and that her sister was experiencing now. Her classmates were reading the *Times* and the *Journal* every day. They came from urbane East Coast families, in the shadow of New York. Margaret's heart was in the outdoors—camping, hiking, climbing, cycling, rafting. Through a friend, Margaret got her first glimpse at the fine arts department. She knew that it had everything she wanted, yet the life of an artist was an extravagance she could not afford.

Margaret was a work-study student, serving and cleaning up after her classmates in the cafeteria. She no longer benefited from the financial cushion of being an adjacent member of the Gary family, and she was starting to realize that the last several years she had lived off the Garys' generosity were, in some ways, an illusion. At the end of her freshman year, Margaret decided to transfer to the University of Colorado in Boulder. CU was cheap enough for her to afford on Pell Grants. She had friends there. And it was still a safe enough distance from home—too far to be a commuter, far enough that she could beg off if she didn't feel like coming home for visits.

Every decision Margaret made was, in some way, oriented around the ability to avoid going home. Home was where Peter was urinating on the floor because a devil was under the house. Home was where Donald was still ranting and raving about his ex-wife, a decade after the divorce. Home was where Matt was cooling off, after his psychotic break at the Garys' house. And home was where Jim was still welcome to drop by anytime he wanted.

In Boulder, Margaret was in classes with many of her old Kent friends, the rich ones who traveled to France or Portugal in the summer. She did her best to have enough money to at least have fun domestically. She scooped at Steve's Ice Cream and had a second, semiregular job dealing mushrooms for a much older supplier—a guy who leered at her a lot but never made a move on her. With an old Kent School boyfriend, she saw as many as fifty Grateful Dead shows, all around the country, most of them while drenched in coke and acid. Margaret wanted to feel strong and capable and independent. But some part of her was waiting to be rescued—to keep her from ever having to engage directly with anything deeper.

> Why do I even go home? My mind feels like it's going to
> wind up so much that it won't ever stop spinning. I cannot
> understand or cope with my brothers, especially Matt,
> Peter, Joe and Donald. I'm in tears right now because I
> can't handle any of it. . . . Life is merely the permanent
> roots your family knots around you. My family depresses
> me, they hinder my progress in many ways. I'm stuck with
> insanities that no one should have to go through life trying
> to ignore. . . .
>
> Margaret's diary, April 3, 1983

That summer, Margaret was out east following the Dead when she found herself swept off her feet in a way she only had dreamed.

Chris had been an upperclassman at Skidmore when Margaret was there and had noticed her then. In college he'd been known as Hot Knives—the name for a technique in which you take a piece of hash and smash it between two red-hot knives, and then inhale the smoke. When Chris saw her again now, at a party in Connecticut, he made his move.

Chris was a few years older than she was, with an aggressive, nimble intellect. His father was an oil executive, and Chris was a fixture at his family's yacht club, racing Laser-class sailboats in championships around the world. He paid to fly Margaret out to Maine in August to see him again. They went boating to the islands off Georgetown and Boothbay, drank Bloody Marys and blueberry daiquiris, ate lobster, and brought nineteen more of them back to Connecticut, where he introduced her to his father. The next day, they sped into Manhattan in his BMW for shopping at Saks Fifth Avenue and Bloomingdale's. To Margaret, Chris wasn't just another guy. He was an entirely new narrative.

> I never thought I'd meet a man with so much to offer, and
> the outrageous part is he wants to share it with me.
>
> August 31, 1983

She went back again in September. He flew out to see her in Colorado in October, and again on Thanksgiving. And on New Year's Eve, they were together again in Manhattan, dressed brilliantly, ringing in the new year at the Rainbow Room. They both were half-done-in by champagne and coke and pot when, in the first moments of 1984, Chris leaned in toward her, conspiratorially.

"Can you keep a secret?"

"Yeah."

"Will you marry me?"

"YOU'RE NOT GETTING married to this guy. That's ridiculous."

This was Wylie, a classmate of Margaret's in Boulder, another love interest, or at least he hoped to be. While Chris was a competitive sailor, Wylie spent the warmer months painting houses. Wylie was level-headed and soft-spoken, usually. But this news, and the ring on Margaret's finger, took him by surprise.

But she was serious. No one was taking care of her anymore—not her family, not the Garys. Chris was ready. Trips to Germany and Crete and Egypt were all planned out.

Lindsay got it. She might have been the only other person on the planet who really knew what Margaret was running from. This was her sister's chance to have a new family.

Mimi and Don approved, too. Aware of Chris's family's wealth, they mortgaged their house to host the finest wedding they could manage. Mimi made all the dresses herself from an Oscar de la Renta pattern, pink silk with ruffles around the bottom and top.

They set a date for August. All Margaret needed to do now was navigate a path through her brothers—all nine of them—to the altar without a scene.

IN THE MONTHS before the wedding, Peter was arrested in Vail for soliciting funds on the street for what he was calling a cancer society benefit. At the hospital, he asked the doctors for a bulletproof vest to protect himself. The Vail police, he said, were jealous of him and out to get him. Eventually Peter made it back to Hidden Valley Road with Mimi and Don, staying in bed, not bathing, subsisting on coffee and cigarettes, alternating between long periods of silence and occasional explosive outbursts. Once, he locked Mimi out of the house and put his medicine in the family's coffee.

Two of the other hockey brothers, Joe and Matt, were in and out of Pueblo at the same time. Joe was preoccupied by Catholic imagery, like his brother Donald, but never grew menacing like Donald once had; the voices in his head were not so much evil, he would say, as bothersome. Matt's fantasies were more paranoid, making it hard for him to stay stable for long. Between hospital stays, he was arrested once for loitering in Colorado Springs and placed on probation.

Donald had been living more or less peaceably at home since his last state hospital visit in 1980. Now the one everyone was most wary of was Jim.

Earlier that year, after sixteen years of marriage, Kathy had finally left her abusive husband. For years, she'd worked and raised their son, Jimmy, while steering around Jim's ups and downs. Her friends all knew about Jim—his mental illness and the abuse—and yet Kathy never made a move until the first time she saw him strike their son.

Jimmy was fourteen. Jim hadn't touched him before then. He saw Jim about to hit Kathy and got between them, facing off against his father for the first time, trying to protect his mother. When Jim punched his own son in the stomach, Kathy called the police. She left with Jimmy soon after.

Now Jim was living alone, still getting outpatient shots of a neuroleptic drug to keep his symptoms in check. But he was working less and drinking more. No one in the family knew what he might be capable of.

A few days before Margaret's wedding, Jim came by the house on Hidden Valley Road, where Lindsay was staying with a boyfriend for the weekend. When Jim arrived, Lindsay wasn't there, but her boyfriend's car was. Others in the house watched Jim as he slashed all four tires, screamed obscenities at the top of his lungs, and drove off.

Everyone who was staying in the house was forced to evacuate so that Jim could not find them. If there had been even a little doubt in Lindsay's mind that Margaret was right to start a new life with Chris, there wasn't any now. Part of her wished she had a similar ticket out.

THE REHEARSAL DINNER was at the Garden of the Gods country club. At least two hundred people would attend the church ceremony, followed by a reception in the backyard of a family friend's new home in Old North End, the fanciest part of Colorado Springs.

Wylie called Margaret the night before the wedding with a last-ditch offer. He was in Massachusetts with his family. "I'll send you a ticket here if you don't marry him," he said.

Margaret cried for hours. Lindsay stuck an ice pack on her face to keep the swelling down. Margaret knew that she wasn't doing the right thing, that she was about to marry a man she hardly knew. But what was the alternative? Fly to Wylie? Cry on his shoulder? Tell him that one of her brothers had molested her for years—and that another killed himself—and that there were four more at home just like them?

To Margaret, that was no choice at all. Wylie wanted more from her than she could give anyone—a sincere, honest look at her own life. With Chris, she wouldn't have to think about her family ever again.

CHAPTER 26

She hadn't counted on missing the mountains so much.

Lindsay graduated Hotchkiss in 1984 in the top quarter of her class. She could have found a college farther away from home than Boulder. But Colorado, she was amazed to realize, had been calling to her—not Hidden Valley Road, exactly, but something about the state that felt like home. Now that she was back, she wanted to climb every fourteener she could see, all the time. And for a short time, she could commune with that place again, until all her usual fears came back.

At the University of Colorado, she got straight As doing hardly any work, and yet at odd moments she was overcome with panic. She had a social life, boyfriends, parties, drugs—nothing was stifling her anxiety. She found herself reading every self-help book she could find at any bookstore, trying to figure out why.

When she tried mushrooms for the first time, she thought that this must be what schizophrenia felt like: absolutely terrifying. She didn't need mushrooms to be afraid. She had plenty to worry about without them.

She grew tired of pretending that nothing was wrong. She was looking for help, but she was unsure of where to find it.

"TELL ME ABOUT your family," the campus therapist said.

Lindsay started talking. And then something happened. As she started explaining that she had ten older brothers and that six of them had schizophrenia, the look on the therapist's face changed.

At first it seemed like she didn't believe Lindsay—that she thought she was making the whole thing up. Then Lindsay saw what was really happening. The therapist was wondering how much of this was all in Lindsay's head. She thought she was the crazy one.

The session went nowhere. Who would listen to her? Who would believe her?

That fall, Lindsay started seeing a boy, someone she'd known for years. Tim Howard was Sam and Nancy Gary's nephew. Like Lindsay, he had been visiting the Garys' lake house in Montana his entire life—another of the many children Sam and Nancy would host. Like a lot of boys, Tim had been in awe of the Galvin sisters—both stunning, both effortlessly athletic. Now he and Lindsay were in college together in Colorado.

Lindsay and Tim had been dating a few months when they both ended up as guests of the Garys in Vail during a school vacation, staying at the family's condominium on the main strip. There came a time when they finally had the place to themselves—everyone else was either skiing or shopping—and they were on the verge of sleeping together.

Lindsay couldn't.

Tim asked her what was the matter.

Lindsay looked at him.

This wasn't an angry boyfriend, demanding sex. This was a boy, nearly a year younger than she was, who had been carrying a torch for her for the better part of a decade—a boy who genuinely liked her, who would not judge her. He knew a little bit about her family already, even if he didn't know some of the more difficult details. And this was Tim, not some stranger. There may have been no safer person to tell.

Lindsay was in tears as she talked. This threw Tim, at first. She had always seemed so tough to him—a *shtarker,* like Sam had often called her; Yiddish for a tough guy, someone who knew how to get things done. But he stayed in the room with her. He listened.

She stopped short of revealing Jim's identity. She didn't say who had abused her, and he didn't ask. When she stopped talking, he struggled with what to say.

"I don't know what to do," Tim finally said. "But I know who would."

They got dressed and left the condo when Tim spotted Nancy Gary in the distance, walking toward them along the main drag. Tim left Lindsay and ran up to his aunt. "Can I talk to you for a minute?"

Lindsay stood there, snow on the ground around her, as Tim and Nancy talked. Barely a moment passed before Nancy cut away from Tim and marched down the lane to Lindsay. She and Nancy went inside and talked some more.

—⁓—

Louise Silvern remembered meeting Lindsay for the first time in 1984, listening to the pretty, self-possessed nineteen-year-old talk about her family and what had happened to her. Lindsay's description of her family, and of the minute-to-minute experience of growing up in that house, was far and away the most traumatic story, certainly, that she had ever heard from a patient. And when Lindsay got to the part about the college health services therapist not believing her, she remembered being outraged. Job one, Silvern had always thought, was to not shut a patient down.

There is a narrative, or a myth, that our society indulges in about trauma and therapy, particularly in the wake of unspeakable childhood abuse. The myth starts with a child unable to speak, and takes flight when the right therapist is sensitive and kind enough to coax the child into a breakthrough. This is the mold established by Dr. Fried, the Frieda Fromm-Reichmann surrogate in *I Never Promised You a Rose Garden.* Once the child lets it all out, the trauma disappears like a bad dream. The patient is as good as cured—relieved and unburdened and ready to embrace the world again. In books and movies, the breakthrough happens in one fraught, angry, tearful session, perhaps late at night, after a small crisis triggers something in the patient that they've tried to keep bottled up for years.

In Lindsay's case, the myth was barely half true. In Silvern— Lindsay's second therapist, based in Boulder and referred by Nancy Gary—Lindsay found a professional listener who, yes, through sensitivity and kindness, created the safe, accepting space that was necessary for Lindsay to take control of her own story.

Where the myth breaks apart is with the idea of a breakthrough. For Lindsay, the breakthrough was more like a seep-through, coming gradually, over twenty-five years, the product of steady, intense work in sessions that sometimes were as frequent as three times a week. While Lindsay was going to classes and getting straight As and having boyfriends and going skiing and climbing, she was dashing away for an hour a week, sometimes two or three, to tell her therapist her family secrets. And while this took a very long time, Silvern made sure the pace remained unrushed. Unlike the movie therapists, she did not want to seem overly invested in the outcome of each session. That kind of pressure can turn a patient into a performing seal, just doing whatever she feels the therapist expects. At its worst, that pressure can be retraumatizing.

As a first step, she did very little but listen to Lindsay carefully for several sessions, paying attention to which subjects were overwhelming, or "fragmenting," to her, and which closed her down entirely. To become fragmented, she explained, was to be so walled off from difficult elements of yourself that those difficulties would only grow stronger, more insistent, more destructive. The solution, or the goal, was to help Lindsay find her own strengths and then develop them to help herself cope with these challenging subjects—to "integrate," as Silvern put it, the difficult parts of her psyche into the rest of her life, rather than cordon them off.

Lindsay wanted to move faster, of course. She wanted to get the problem solved—for someone, anyone, to send the worry away. But for her brothers' and her own sake, she also wanted answers from Silvern about the nature of mental illness—the causes. Could trauma or abuse cause insanity? Is it possible that Peter or Joe or Matt were at Pueblo because of something Jim did to them?

It seemed like a tidy enough explanation. But if that were true—and to be sure, no studies have ever suggested that abuse does cause schizophrenia—that would mean that Lindsay was at risk.

After all this time, she still was terrified of becoming mentally ill. Silvern made it clear to Lindsay how much bravery it would take for her to get past this fear.

———

LINDSAY PAID FOR the sessions herself. Silvern would put whatever she couldn't pay on a tab. Lindsay continued to pay it off for years after graduation, settling it finally after starting her own business in her late twenties.

She never asked her parents to pay. Both Mimi and Don rejected the whole idea of therapy. *Why dig all that up again? Let the past be the past.* Exactly the response that made Lindsay ashamed in the first place, afraid to tell them the truth about what Jim did to her.

Silvern focused on getting Lindsay to tell her own story—to reclaim the past on her own terms. This was about more than just trying to face up to reality. It was about removing all of the filters that had been imposed on her. Children, Silvern explained, rely on the adults around them to interpret what's happening to them. They use their parents' constructed systems: This is good and that is bad; this person is untrustworthy, and that person is somebody you can count on. Shame and guilt are ways that children usually process those traumas when the grown-ups around them have failed them.

Exhibit A for Lindsay, of course, was Jim.

Jim was still in all of their lives, a member of the Galvin family in full standing, turning up on holidays, popping by Hidden Valley Road whenever Lindsay visited, even living back home for a time after Kathy left him. Now that she was back in Colorado, Lindsay was working hard to make herself okay with that, showing up at events like Margaret's wedding as if everything was fine. But Jim was only getting more volatile, now that his wife and son were out of the picture. And Lindsay was getting tired of pretending.

Lindsay asked her therapist: *How can I be around him? How can I go home, knowing he'll drop in at any moment? And if I refuse to come home, can I deal with the upset that would create?*

Silvern would help Lindsay fantasize about what she could do with her anger toward Jim. Lindsay thought about killing him—a lot—and then she felt guilty for those thoughts. But her biggest concern, even bigger than confronting Jim, was that she would have to tell her mother. What if Mimi didn't believe her? *Then,* she thought, *I would somehow be another crazy one.*

She was stuck in the same dilemma she experienced as a little girl: If you were angry, you were unstable. If you cried because you got a B on a test, maybe it was time for you to go to Pueblo.

Lindsay's father remained idealized for her—in her mind, at least, her only ally left on Hidden Valley Road, despite his frailty. But she and Silvern talked a lot about the particular way Mimi had of silencing Lindsay. She wouldn't say, "Shut up." It was more like "You think you've got troubles?" She attacked Lindsay's emotions by undermining them, dismissing them, or invalidating them.

Feelings were scary in the Galvin family, Silvern said. There had been too many out-of-control horrors for it to be otherwise.

SILVERN CALLED RESILIENCE "that wonderful term for something we don't understand." Resilience is the subject of umpteen academic studies, of course, and if someone could figure it out, they would rush to bottle the solution. In Silvern's experience, it was sometimes a matter of luck that a person has the right temperament to absorb trauma in a way that still allows them to be open to new experiences, to go through life with armor.

But there are all sorts of coping mechanisms, some more self-limiting than others. Lindsay was a tough kid, donning a mask of self-reliance and stubbornness that served her well through childhood, and then eventually that mask fused to her real face. The question was how well that mask was still working for her now: hypervigilant, uncomfortable with failure, terrified to present herself to others as anything less than perfect.

Silvern told Lindsay that when somebody copes by being more armored, it can wind up hindering them later. They have a narrower road to travel going forward—a more fenced-in, claustrophobic life. Her hope for Lindsay was that she end up in a place where she would be willing to trust new people, to let down her armor under the right circumstances.

To get there, Lindsay would have to learn to recognize post-traumatic stress in real time, as it was happening to her—so that she would be able to recognize, for example, that a blistering argument she had with a friend one night was at least in part because of the rape scene in the movie they'd just seen.

THERE CAME A time in her sessions when Lindsay decided to talk about what had happened to her at the party in eighth grade, the night in the closet. She was vague about it at first—"there was an incident with some boys."

Silvern knew that Lindsay had to go at her own pace. First, she needed to work through all the self-blame.

She lied to her mother. She went to a party when she shouldn't have. Didn't she deserve what happened next?

Come on, Silvern said. No.

Was she asking to be taken advantage of?

No, Silvern said.

Was she sending out some sort of sexual signal, as the victim of her brother's abuse, equating sex with affection in some misplaced way? Was she asking for it?

No, of course not.

Why didn't she just leave the closet?

Because there were three boys in there with her.

And then Silvern took a risk and used the word.

"They raped you," Silvern said.

Lindsay was not scandalized. She was, if anything, relieved. Someone was giving what happened a name.

The defining of terms was like a glass of cold water splashed on her face. Sexual abuse was sexual abuse. Rape was rape. Being a victim was being a victim. She couldn't escape the closet that night for the same reason that she couldn't leave Jim's cabin at the top of the Manitou Incline: Because someone more powerful was violating her trust, victimizing her, making it impossible for her to do anything other than what he wanted her to do.

Next came the careful unpacking of the details. Where the particulars of every incident were once completely off the table, now reciting all that dreadful minutiae was helping Lindsay regain a sense of control. The details reinforced how unrealistic some of her self-blaming notions had been. (Unrealistic, yet understandable—children usually have no way of processing trauma beyond their own experience, and so, all too often, they blame themselves.) And to articulate all of that in front of Silvern—seeing how it was possible for somebody who really cared about her to still see her strengths and respect her and know who she was, even though they knew everything about what she'd been through—was a first for Lindsay. In a way that no one in her family ever could, Silvern gave Lindsay a place where she could own her own emotions and express them on a regular basis.

Talking with her therapist about being raped by those boys was, in itself, a tremendous step for Lindsay to take. It also was a perfect dress rehearsal for what had to come next. She would have to be just as transparent with her family as she had been with her therapist. Only this time, the subject would be Jim.

—⁓—

They were in the car, Lindsay and her mother, going to Mimi's friend Eleanor Griffith's house. They pulled up to the house. They parked and walked slowly to the entrance. They saw that Eleanor was not home yet.

They were alone, mother and daughter, with a stolen moment. This was when she chose to talk about it.

Lindsay had already been opening up more to Mimi, writing her long, philosophical letters from college about the family and the illness. She wrote about what it was like to grow up around Donald, and how no one acknowledged the pain that caused her. She wrote about the state of fear she inhabited in those years. Mimi's reaction was always the same. She would acknowledge what her daughter was saying and then urge her to move on—to forgive—always reminding her that there was someone else out there who had it worse. It was superb maternal jujitsu: paying lip service to relating to her daughter's experience when in fact she was obliterating it, draining it of all meaning, blotting it out.

So it shouldn't have surprised Lindsay when, standing there in front of the Griffiths' house, she started to tell her mother that she had been sexually abused by her brother Jim countless times over several years—and her mother responded by saying that when she was a girl, the same thing had happened to her.

IN THE OFFICIAL version of Mimi's enchanted New York City childhood—the story she'd raised her daughters on, and related to friends and neighbors proudly—Mimi's stepfather, the painter Ben Skolnick, was her tutor in music and art. While her mother worked in the garment business in Manhattan, her stepfather helped her appreciate culture in a way no one ever had before. All of that was true. He played Tchaikovsky for her on the record player. When she was laid up with a sprained ankle, he suggested *Carmen*.

But it was also true that Ben drank, and it was also true that he took liberties with Mimi. When Lord & Taylor started selling Mimi's mother's A-line skirts, she couldn't manufacture them fast enough, and she started spending most weeknights in the city—leaving Mimi at home with her stepfather. That was when Ben Skolnick advanced on her.

Mimi was deliberately light on details, and Lindsay did not press her for any. But it was clear that he'd molested her, touching her inappropriately.

As she told Lindsay this, Lindsay sensed some of the stray threads of her mother's childhood story coming together. She understood now why the marriage between Mimi's mother and Ben did not last—why they lived apart after the war. And Mimi said one thing that, in an instant, made Lindsay think of her mother entirely differently. Mimi said that she finally told her mother about it when Ben started to prey on her little sister, Betty.

Lindsay knew something about the nerve it would have taken a girl in that position to speak out—to put her own credibility on the line to save her sister. If her mother had really done that, then Lindsay must not know her as well as she thought she did.

That exchange with Mimi might have been the most emotionally complicated moment in Lindsay's life. Part of her was knocked flat by her mother's candor, and after hearing her mother's story she felt closer to her than ever. But at the same time, Lindsay felt she had been denied something—her own misfortune was once again preempted by someone else's. Mimi was talking about her own experience, skipping right past the details of what Lindsay was saying about Jim. Lindsay needed Mimi to take her side, to tell her that what Jim had done to her was wrong.

But Mimi did not do that. She had never picked the side of a healthy child against a sick one, and she wasn't going to start now. Instead, Mimi started talking about how Jim was mentally ill.

Lindsay flushed. To her, schizophrenia wasn't an excuse for what

Jim had done to her. Certainly no mainstream researcher or psychiatrist would say that it was Jim's psychotic delusions that made him a pedophile.

But Mimi was not willing to separate the two issues. Lindsay, though she expected as much, was still deeply hurt. What made it so hard for her mother to sympathize with anyone other than her boys? It was as if she had used up all of her compassion on the sick children, even Jim, leaving nothing for anyone else.

But that day, Lindsay was ready. She told her mother she would never agree to be in the same room as her brother again.

—⁂—

Jim wasn't supposed to be there. Her parents had assured her he would not be.

Lindsay was back on Hidden Valley Road, visiting for a Sunday dinner after a long absence—her first time back since that night outside the Griffiths' house. Both of her parents were there. So was Joe, medicated and somber and, unlike his other sick brothers, acutely aware of his own sickness. A peaceful evening for the Galvins, until Jim walked in.

Her father asked him to leave at once. "Jim, you don't belong here, please go home."

"Why don't I belong here?" Jim said.

Mimi said nothing.

Lindsay bit her lip. It didn't help. She lost it. She stood up and started screaming.

"You fucking asshole! You sexually abused me!"

Jim was not in good shape. His wife and son had left him, he was heavily medicated—and, per one of the side effects of the medication, well on his way to becoming obese. But he was not conceding anything, and he was more than willing to retaliate. He picked up a guitar that was lying around and broke it in half. He called Lindsay a liar, and he started yelling and screaming.

"That's not true! You're imagining things!"

But Jim could read the room. He saw no one was listening to him. And then he saw his father, telling him to get out and that he never wanted to see him there again.

Jim left. Lindsay spent the rest of the evening in tears. Her parents left her alone, heading back to the kitchen to do the dishes. Joe comforted her. "You're not lying," he said, holding her. "I know you're not lying."

That was what Lindsay would think of most in the years that followed—how her brother Joe believed her, and how her father had, too.

CHAPTER 27

New imaging equipment—including CAT and PET scanners—has demonstrated physiological differences in the brains of some schizophrenia patients. And now, using this and other technology, NIMH, under staff psychiatrist Lynn DeLisi, is attempting to identify a genetic marker in families where more than one member suffers from schizophrenia. . . .

Both healthy and ill family members are needed for the study. Patients will continue to be treated by their regular physicians and participants will be paid.

Prospective participants may call Dr. Lynn DeLisi: 496-3465.

The Washington Post, July 20, 1984

From her seat at the long handmade wooden table in the kitchen on Hidden Valley Road, Lynn DeLisi saw at once the burden that Mimi had been bearing all these years.

Her husband was home and frail. He could help around the house and even drive, but each night he would go to bed wondering if he would remember what he'd read the next day.

Donald, the oldest, was home more or less all the time now, too.

Three other sick sons, Joe, Peter, and Matt, roamed in and out of the house, between the hospital and home and their own apartments, which they'd inevitably leave or get tossed out of. Even Jim would wander by from time to time, too, before Don would notice him and demand that he leave.

The violence was a little less frequent these days. They were all getting older now, and they all were more consistently medicated. It was up to Mimi to keep them all active, manage their care, shuttle them to appointments, dispense their meds.

Given all that, DeLisi was amazed by the good cheer the Galvin matriarch displayed. "You can't be heartbroken *every* day," Mimi liked to say.

IN THE YEARS before Lynn DeLisi walked through the Galvins' door, there was still no single theory of schizophrenia that was universally accepted. The precise mechanism of the disease remained a mystery, and many of the same nature-nurture battles continued. But without any great fanfare, some things were slowly changing.

After three decades, the schizophrenogenic mother theory was losing its hold. In 1982, an Australian psychiatrist named Gordon Parker published a review of schizophrenogenic-mother research in *The Journal of Nervous and Mental Disease,* concluding that, while distant and controlling mothers no doubt existed, there was no evidence that they were more likely than anyone else to have children with schizophrenia. The next year, Chestnut Lodge—the institution that, under the direction of Frieda Fromm-Reichmann, steadfastly ignored all calls to treat schizophrenia as a biological disorder—experienced a dramatic reversal. Thomas McGlashan, who had joined the hospital as a therapist in the 1970s, went public with a study of the case records of every patient treated there between 1950 and 1975. His conclusion: Only one third of Chestnut Lodge's patients were moderately improved or recovered. If, like the psychoanalysts of Chestnut Lodge, you believed that the right course of therapy could cure almost any psychotic patient, a 33 percent success rate was not something to be proud of—especially not when the pharmaceutical industry was claiming a much higher success rate in treating the symptoms of psychosis. "Frieda . . . embarked on a grand experiment," McGlashan said at the time. "The data is in. The experiment failed."

After decades of debate, the thinking about schizophrenia seemed to be consolidating around the physical nature of the disease. On *The Phil Donahue Show* in 1983, the NIMH psychiatrist E. Fuller Torrey, promoting the publication of his book *Surviving Schizophrenia,* one of the most popular and influential books on the illness that decade and beyond, showed the audience images of CT scans of healthy brains, contrasted with brains with enlarged ventricles of schizophrenia patients. "That's the brain disease you are looking at," Torrey said. In a study published that same year, Torrey and his colleagues on Richard Wyatt's team had ruled out neuroleptic medications as the cause of these larger ventricles; it was the illness, not the medication, that seemed to create this difference. Anyone who could not acknowledge now that schizophrenia was physical, he joked, must be a little behind in their reading. "Unfortunately there is a segment of the psychiatric community that reads only the *National Geographic,*" Torrey said. "They have not got the word yet."

This was the age of biological psychiatry now, with psychopharmacology not far behind. The latest DSM—the DSM-III, published in 1980—had narrowed the diagnostic criteria of schizophrenia to seem less like the syndrome it was and more closely resemble a specific illness. Based on this new criteria, even Joanne Greenberg, the author of *I Never Promised You a Rose Garden,* was said to have been misdiagnosed at Chestnut Lodge. The delusional teenage girl did not have schizophrenia at all, a team of researchers declared in 1981, but merely suffered from an episode of somatization disorder, once known as hysteria—fleeting hallucinations coupled with acute but temporary physical pain. Schizophrenia's star patient might not have been that sick to begin with.

It was a little too early, however, to declare victory in the nature-nurture war. With talk therapy on the ropes, neuroleptic drugs were ascendant. These drugs changed the lives of thousands of people, helping them create some space between themselves and their delusions. In the popular imagination, and even among many doctors, neuroleptics were considered revelatory, like insulin for diabetes. But how could that be when schizophrenia itself remained ragingly mysterious, and the drugs themselves could be physically damaging? The drugs made some patients obese, others stiff and ungainly, others practically catatonic—this from drugs that had been hailed as miracles. For the chronically mentally ill, success had been defined down to a point where it was starting to look a lot like failure.

The only real, unambiguous beneficiary of drugs, of course, were pharmaceutical companies—all of which were still developing variations of the same original drug, Thorazine, that had been developed back in the 1950s. Then again, their very efficacy had seemed to stifle innovation. Why was it that every new drug brought to market had been either a version of neuroleptics like Thorazine or atypical neuroleptics like clozapine—with no disrupting third class of drug to spur forward progress?

For the first time, large numbers of families of the mentally ill were speaking up, forming advocacy organizations and a patient's rights movement, trying to get across how their struggling daughters, sons, brothers, sisters, wives, and husbands felt betwixt and between—unreached by traditional psychotherapy, yet only pacified by drugs. For the many patients who felt ill-served by drug therapy, the decision to treat schizophrenia as a physical illness had yoked them to a treatment that held no hope of a cure. Their dilemma was real, and painful, with no clear answer. Those who rejected the pharmaceutical therapies argued, just as R. D. Laing and others in the anti-psychiatry movement had in the 1960s, that not every society anesthetizes its unconventional thinkers. But for most people with a loved one diagnosed with schizophrenia, it was almost impossible to witness what they were going through and see anything other than suffering—and even harder to think of what, besides powerful drugs, might help.

Until the illness could be understood better—the code of schizophrenia cracked, and a proper therapy produced that might lead to a cure—these patients, including the Galvins, were, sadly, a captive market.

DELISI BEGAN COLLECTING the genetic material of families with schizophrenia as a researcher in Elliot Gershon's lab in 1984, almost a decade after her tentative first days at NIMH. What once seemed impossible in her early years there was now tantalizingly within reach. Advances in molecular biology now made it easier to quickly copy a piece of DNA thousands of millions of times—allowing the genetic code, once the great unexplored realm of human biology, to be deeply probed for the first time. With these new tools, researchers elsewhere already had isolated the gene for one disease: phenylketonuria, or PKU, which caused intellectual disabilities. Others were going after Huntington's disease. But those illnesses were a far cry from schizophrenia, which almost everyone agreed had to be the work of not just one mangled gene but many. A disease as complex as schizophrenia probably had a genetic makeup no one could completely see with the tools available at the time. The thought of traveling the country, collecting DNA from families, struck many of her colleagues in other labs at NIMH as a fool's errand.

But DeLisi was as sure as ever that multiplex families held the answers. She didn't mind if others thought of her as out on a limb. "Lynn would think along lines other people wouldn't," Gershon remembered. "She could go in different directions."

She found her first family without having to leave the hospital. A patient Gershon had been treating in his clinical practice happened to have a brother who had also been diagnosed with schizophrenia. DeLisi learned that the brothers' parents, Jim and Carol Howe, had been among the founders of the National Alliance for the Mentally Ill (now known as the National Alliance on Mental Illness), an advocacy organization that started in 1979 in Wisconsin and was expanding with new branches around the country. If DeLisi wanted to find families quickly, she thought, NAMI would be the perfect ally.

DeLisi contacted regional chapters of NAMI and asked them to advertise her study in their newsletters. The families that came forward generally had two or three people with schizophrenia; one or two families had as many as four. As more responded, DeLisi hired a social worker to visit families she could not meet with personally. But when she heard about the Galvin family of Colorado Springs, DeLisi knew she had to fly there and see them herself.

As she walked through the door of the house at Hidden Valley Road, she couldn't help but recognize a perfect sample. This could be the most mentally ill family in America.

DELISI ASKED EVERYONE in the Galvin family, even those who were not diagnosed mentally ill, to participate in psychiatric interviews to confirm or rule out a diagnosis for each of them. Then she drew blood samples in hopes of noticing something in this family's genetic makeup that might indicate a propensity toward mental illness. Some family members might be carriers who did not get ill, she believed; the markers could be present in everyone.

All the sick brothers participated without much of a fuss; Mimi had made DeLisi's work easier in the way she had always closely supervised the care of all of her sick sons. Among the six well siblings, everyone agreed except for Richard—the sixth son, once the teenage schemer, now a mining investor in Denver—who was still too unnerved by the family illness to engage in any of his brothers' treatments. (John, the third son, now a music teacher in Idaho, had his blood drawn remotely and sent to DeLisi's lab.)

Lindsay and Margaret came away feeling hopeful that the research might lead, someday, to a breakthrough. The look on Mimi's face, meanwhile, was practically beatific. The most important breakthrough, in her view, had already happened. She had been waiting for decades for someone like Lynn DeLisi to come knocking on her door. Now she was finally here.

ROBERT FREEDMAN'S FIRST visit to Hidden Valley Road took place very soon after Lynn DeLisi's. On that day—and on subsequent visits by various Galvins over many years to Freedman's lab in Denver—Freedman and his team from the University of Colorado Medical Center's psychiatric research division recorded the Galvins' brain waves, drew their blood, and administered questionnaires. As he got to know the family, Freedman marveled at how Mimi kept the boys at home much longer than many families would have. "She was delightful," he said.

Freedman was stunned by the decision to send one of the daughters, Margaret, away to live with another family. How horrible things must have been at home, he thought, for Mimi and Don to even entertain such a drastic decision. He saw that Don's health was declining and that the sick boys were a handful. But above all, he was struck, as DeLisi had been, by Mimi's determination to care for them all. "Medications in those days made the boys very stiff and unresponsive. So they kind of sat there like hunks and they weren't talkative, and she was left to manage them. She was running a rooming house."

DeLisi had tipped Freedman off to the Galvins, knowing that he had been looking for families to test his sensory gating theory.

Freedman had spent the early 1980s running his double-click studies, designed to measure the brain's ability to filter information. He continued to believe that sensory gating was a mechanism in the brain, something genetic that made certain people susceptible to schizophrenia. And he felt as if he was getting warmer. In 1984, just before meeting the Galvins, he had studied the gating abilities of schizophrenia patients and members of their immediate families, and he found that half of the immediate family members had the same gating deficits as the family members diagnosed with schizophrenia. Here was another sign that he was on the right track—evidence that sensory gating was hereditary.

Why some siblings with sensory gating issues ended up manifesting the symptoms of schizophrenia and others did not was still a mystery. Freedman's next step was to try to locate the specific part of the brain responsible for sensory gating. Thanks to DeLisi, he now had access to a family with an unfathomably, overwhelmingly profound manifestation of schizophrenia.

IN FEBRUARY 1986, months after her first visit with the Galvins, DeLisi used data from her families to confirm what Richard Wyatt's NIMH team had discovered about schizophrenia's correlation to large brain ventricle size. A year later, she used the data in a study testing a possible link between schizophrenia and human leukocyte antigens, or HLA, a gene complex involved in the regulation of the immune system. No such link was proven. Still, the multiplex family database had begun contributing to the body of knowledge about the disease. As far as DeLisi was concerned, this was only the beginning.

She sent the Galvins' blood samples to the Coriell Institute for Medical Research, a facility in Camden, New Jersey, that preserves huge collections of cell lines from various diseases. This allowed for the possibility of others using the family's DNA as a resource in dozens, even hundreds of future studies, conducted in labs around the world. DeLisi held fast to her belief that if she could find a marker for schizophrenia embedded in the genetic data of a family like the Galvins, schizophrenia might one day become like heart disease, an illness with particular benchmarks and risk factors that could be measured. In 1987, DeLisi was recruited away from NIMH by the State University of New York at Stony Brook, which offered her a professorship and a program of her own to run. She kept researching multiplex families there. She had forty already, including the Galvins. With a grant from NIMH, she steadily built on that list, eventually reaching one thousand families—more than anyone else had managed to assemble.

Then came several fallow years. Family studies were yielding amazing results in other diseases, including early onset breast cancer and Alzheimer's disease, but there was no breakthrough for schizophrenia. In 1995, DeLisi published two studies drawn from her own pool of data on families. The first seemed to confirm that the same genes responsible for schizophrenia are connected to other mental illnesses like depression or schizoaffective disorder. The second failed to find a link between schizophrenia and bipolar illness, at least on one particular chromosome where bipolar illness appeared to be rooted. DeLisi remained confident that someone somewhere could find a genetic fingerprint in this pool—and show that nature, not nurture, determined this condition. "I am not a firm believer in environment having an effect at all," DeLisi told a reporter in 1999.

DeLisi's work still had supporters. "It is critical that we avoid premature disillusionment," Kenneth Kendler, with the Medical College of Virginia, wrote in 1993. "The human brain is very complex and quite difficult to access." But one of her old colleagues from Richard Wyatt's lab at NIMH, Daniel Weinberger, started to suspect that researching families was a blind alley. "More than ninety percent of the relatives of schizophrenics do not have schizophrenia according to current diagnostic criteria," he'd told a reporter in 1987.

Weinberger had a point. The odds of siblings in the same family sharing the condition are indeed low. On the other hand, a sibling of someone with schizophrenia still had about ten times the chance of having the condition as a person in a family without the disease. Compared to the odds of inheriting many other disorders, these odds were extraordinarily high—higher, even, than heart disease or diabetes. Seen that way, it would seem foolish *not* to keep looking at families.

AT NIMH, THE search for more physical signs of schizophrenia continued, even as the direction of that research seemed almost aimless. Wyatt's lab had used MRIs to examine identical twins with one sibling with schizophrenia, comparing the size of each twin's hippocampus. Sure enough, in 1990, they found differences. The hippocampi of the brains of people with schizophrenia were smaller than those without. This finding, like the one about enlarged brain ventricles a decade earlier, seemed to reveal something new about how the disease worked: The hippocampus helps remind you of where you are at any given moment, and it is less developed in the twins that, diagnosed with schizophrenia, have less of a grip on reality.

"We were high as a kite on this stuff," remembered Daniel Weinberger, who coauthored both of those studies. "But there was a gnawing feeling in the back of my head." All that this brain research was doing, he thought, was confirming different versions of the same idea: that a schizophrenic brain is physically different from a normal brain. For those who treated schizophrenia patients on a daily basis, this was hardly surprising. "You could talk to these people for five minutes," Weinberger said, "and you knew their brains couldn't be functioning the same."

The MRI studies were seeming less valuable over time—all just pieces of one little corner of a much larger puzzle. Weinberger suspected that the only reason researchers loved them so much was that they had the tools to do them. "One of the things that has characterized psychiatry research forever is the old saying of, 'Looking for the lost keys where the light is.' Everything has been, 'Well, we have this tool. We have a hammer, so we're going to look for nails.' And we would find things, because this is the nature of phenomenology—you find things." Whether they were promising leads or red herrings, no one knew for sure.

In 1987, Weinberger published a theory that went on to change how practically every researcher thought of the illness. Until then, schizophrenia researchers had been fixated on post-adolescence as the moment schizophrenia appears. Brain scans all but confirmed that: The frontal lobe is the last part of the human brain to mature, at the end of adolescence, and MRI studies of the brains of many schizophrenia patients show problems with activity in the frontal lobe. But with his new theory, Weinberger suggested the problems in the brain quietly started much earlier in life. He reframed the conception of schizophrenia as a "developmental disorder," in which abnormalities that patients possessed at birth, or even in utero, set off a chain of events that, in essence, sent their brains off the rails gradually, over time. All genes did, he said, was establish a blueprint for brain development and function. The rest happened later, in real time, with the help of the environment.

If Weinberger was right, the adolescent phase of brain maturation was simply the final chapter of the story. The brain is having difficulties throughout gestation and birth and childhood, only no one notices anything until the final phase of construction, when the brain is mature. Seen this way, schizophrenia's onset seemed a little like a bowling ball that veers ever so slightly to the left or right the second it leaves the bowler's hand and strikes the wood on the lane. For a few feet, the ball seems to be doing well, heading straight. Only closer to the pins does it become clear that the ball has been gradually going off course—so far off-center that it hits just one pin on the side, or falls into the gutter. Back in 1957, Conrad Waddington of the University of Edinburgh had proposed a similar metaphor for explaining the varied directions cells take as they develop and multiply. He envisioned a bunch of marbles rolling down a slope—an obstacle course with an elaborate system of lumps and grooves. Each marble ends up taking a different journey down the slope. That slope is what he called the "epigenetic landscape"—part architecture, part chance.*

This idea made intuitive sense to Weinberger. For all of us, adolescence is a crucial period of housecleaning for brains that had been hard at work for more than a decade of extreme expansion and renovation. This demanding phase for the developing brain explains, for instance, why teenagers need more sleep, or why, after adolescence, it's harder for most people to learn a language or recover from brain injuries. It may only stand to reason, then, that if one's genes lay out merely a potential to develop schizophrenia, this would be when that potential is fulfilled. If nothing else, Weinberger's developmental hypothesis explained why, for example, if one member of a pair of identical twins has schizophrenia, the chance that the twin also will have the condition is about 50 percent—but they each still have an equal

_____

* Waddington's 1957 "epigenetic landscape" model, while famous in its own right, shouldn't be confused with the more recent use of the term *epigenetics,* or the idea of genes activated by the environment.

chance of passing on the disease to the future generations. "The risk is passed on," Trinity College geneticist Kevin Mitchell has written, "regardless of whether the person actually developed the condition."

Whether you get the disease or not, it seems, depends on what happens once the bowling ball hits the lane.

In the years to come, as genetics research grew in scope and ambition, the developmental hypothesis caught on with other scientists. To effectively fight this illness, this theory suggested, one might have to treat people before they seem sick. That, it seemed at the time, would call for isolating the genetic makeup of schizophrenia. Others were joining DeLisi and Freedman in the search for genetic mutations that, in their way, might tell the whole story at last.

DON
MIMI

DONALD
JIM
JOHN
MICHAEL
RICHARD
JOE
MARK
MATT
PETER
MARGARET
LINDSAY

CHAPTER 28

By the time the researchers from NIMH and Denver came to Hidden Valley Road, Donald had become wordless and vacant, his weight increasing, his movements stiff. He had more or less given up on finding a job or even walking around the neighborhood the way he used to. Except for mealtimes, he was a hermit. As painful as this was for Mimi to see, having Donald at home was also helpful to her, in both mundane and profound ways: He accompanied her as she went grocery shopping and did her chores, and he gave her a purpose.

Donald managed to stay out of Pueblo for seven years, instead paying regular visits to Pikes Peak for doses of Mellaril, an antipsychotic, and Lithobid, an extended-release lithium drug that targets mania. Every so often, he would try living in a boardinghouse, but would never last long. It was during one of those stays, around Christmastime of 1986, that he decompensated completely. He was admitted to Pueblo for the eighth time in January, refusing to answer any questions about his marital status (the failed marriage to Jean still loomed large, perhaps) and preaching from the Bible. In a new development, he also was talking about how certain Lithuanians were looking for him and trying to harm him.

Donald told the staff that he had stopped his medications because his watch stopped. Asked about his mother, he referred to her as "my father's wife." Mimi, he had decided, was not really his mother because he was swapped in the hospital—the offspring of an octopus. Pressed to explain his relationship with his family, Donald talked about arguing with his parents about getting a car. Asked if he had a driver's license, he said he had a "Goldilocks and Three Bears" Colorado driver's license.

In a few weeks, he was stabilized on new meds and returned home to Mimi and Don. In the early spring of 1990, after several years of living largely quietly in his room, Donald heard that Peter, after a few failed tries at living on his own, might be coming back to live on Hidden Valley Road. Donald thought that Peter was going to lay claim to his room and decided to take action. He placed phone calls to the Army and Air Force, asking them to station him in Greenland. He announced that he would rather eat in his room than the kitchen; then he went to the market and bought raw octopus and brought it back to his room, leaving it to rot. That was when Mimi noticed that Donald had been missing appointments for his shots of Haldol Decanoate. When he refused to take his twice daily dose of Kemadrin, his parents sent him back to Pueblo.

"My family and I just broke up over financial problems," Donald announced upon his arrival at the state hospital. "I don't want to live in the same house with Peter."

JIM WAS LIVING on his own, getting by on Prolixin. To those who caught glimpses of him, he seemed to be suffering from depression—defanged by years of neuroleptic drugs, obese and frail. His heart was weak, his chest aching with each breath, and yet his paranoia and delusions never completely went away. While Jim was all but an outcast now, his mother still would see him. After everything, he was her son, and she never could shut the door all the way on any of them. The girls never asked about him, and she would try not to bring him up in conversation.

Of all the sick brothers, Joe was the one Margaret and Lindsay found the most poignant in his suffering. Living with Matt for a while, and then in his own federally funded Section 8 apartment, Joe knew that he saw things that weren't there. He went on about Chinese history, and how he had lived in China in a previous life, even as he recognized how strange that was. Once, he pointed at the sky with excitement and told Lindsay that the clouds were pink, and there was a Chinese emperor speaking to him from his past life. "I'm having a hallucination," he said, still half believing it. "Don't you see it?"

Joe was well enough to live alone in an apartment in Colorado Springs, but not quite well enough to fend for himself. When his health benefits couldn't cover his expenses, he piled up too much credit card debt for him ever to climb out of. He filed for bankruptcy with Michael's help. Michael told him he couldn't get another credit card, but Joe got one anyway. He said he had to have a credit card with the Broncos logo on it. Once lean and handsome, he put on a tremendous amount of weight, and his obesity made every little problem worse. His eyesight failed; he developed borderline diabetes. Then came some of the same problems Jim had: chest pain, delirium, stress, panic. But Joe still had his sense of humor, or some of it. He talked with Michael about Transcendental Meditation all the time, hatching plans to try to go to India. He was, in some small ways, still himself. "He had that ability to kind of separate somehow," Lindsay said. "He was the one that would be like, 'I just want this to stop.'"

Joe never stopped wanting a connection to his family, sending religious-themed birthday cards and spending money he couldn't spare on presents. Once, one of Michael's grown daughters was complaining about needing money for books for college, and at Christmas an envelope showed up in her mailbox. Inside was five hundred dollars, with a note saying "for books." Everyone agreed: No one other than Joe would have done such a thing.

ACCORDING TO MATT, the former potter and second youngest of the four hockey brothers, his life took a wayward turn the day his mother decided to send him to a psychiatrist, after his teenage breakdown at the Garys' house. "She took me to CU Medical Center in 1977," he once said. "They put me on a psych ward, but that doesn't make me mentally ill."

Matt was getting more grizzled in middle age, heavier like Jim and Joe and Donald, but also hairier, with a bushy beard, and gruffer, with the imposing bearing of a Hells Angel. Matt's best friends were Vietnam vets and homeless guys who, like him, subsisted on Social Security payments and Section 8 housing vouchers. Matt's doctors learned that he sold his medications on the street more often than he took them.

He was in and out of Pueblo until 1986, when the doctors switched him over to clozapine. After practically his first dose, Matt noticed a difference. He started attending all his mental health appointments without fail. He told his family that he felt like he'd awoken from a nightmare. He no longer thought he was Paul McCartney. An atypical neuroleptic that worked slightly differently from typical neuroleptics like Thorazine, clozapine also had proven helpful to Donald and Joe, but had seemingly little effect on Peter. "When it works," said Albert Singleton, the medical director at Pueblo, "the difference between clozapine and other drugs is like the difference between Bayer aspirin and Oxycontin."

As long as he had a car to drive, Matt filled his day running errands for his friends. He felt useful this way, and he was. He volunteered at a food kitchen for homeless veterans for years; many of the people he served were his friends. The VA sent him a letter of thanks once for his service. "His little bit of responsibility to other people keeps him going," his brother Michael once observed. "I think that's true for all of us."

Even on a better drug, Matt could still descend into long jags of self-pity, airing grievances against everyone in the family and the government. At his monthly sessions at Pikes Peak Mental Health Center in Colorado Springs, he did his best to convince the doctors that he did not need the medication anymore. Every month, he was disappointed. But unlike Peter, who would just stop taking his medicine, Matt's chief reaction was to complain—convinced the entire world was conspiring against him, and that his family had forsaken him. Only when he was at his angriest would his grip on reality loosen a little again. That was when he would became convinced that his medical treatment was not only not necessary, but that it was the cause of any number of world events.

"The more they drug me, the more people will end up dead," Matt once said. "If you ever watch the news lately, like, four hundred eighty people died in four different plane wrecks. Eight thousand people died in an earthquake in the Himalayas; a hundred fifty men in Nigeria got shot down; twenty-two people killed in a church; twenty-two killed in a plane wreck. Quit drugging me, or these things will keep happening."

"I AM THE prophet you have heard so much about!"

In November 1985, Peter Galvin—twenty-five and rail-thin, his hockey bulk a thing of the past—was noticed praying in the middle of a street in downtown Colorado Springs. A few days later, the police ran across him, this time upset and hostile. When they told him he most likely was heading to the state hospital, Peter lost his temper, threatening to fight anyone who tried to take him. When one cop approached him, Peter said he'd rip out his carotid artery. Then he attacked.

This would be Peter's eighth admission to Pueblo. He arrived angry, and at mealtime he refused to eat. During observation, Peter's contradictions became well known to the staff. "It is interesting to watch him function," one psychiatrist wrote. "He says that he will take his medication, but, when confronted that he has recently refused, he says, 'You're right, I have,' as though this inconsistency means nothing to him."

The doctors at Pueblo decided to send Peter to court to face the assault charge as soon as he stabilized. Until then, he was placed in CARES House, the supervised residence that years earlier had given him the boot. Peter tried to escape, climbing out of the same window four times in four days and coming home to Hidden Valley Road. Mimi drove him back each time, only to see him walking back through her door the next day.

After years of questioning his diagnosis, his doctors had finally started prescribing lithium, on the theory that his symptoms lined up more closely with bipolar disorder. But lithium only works if you take it. Mimi and Don were already struggling to referee Peter's clashes with Donald. Now, Peter also refused to take both lithium and Prolixin, and, according to a communication from Mimi included in one medical report, was "not eating or drinking, staying in bed, not talking, staring at family members, being totally unresponsive with occasional explosive outbursts." Her conclusion: "The mother feels that Peter is trying to starve himself and has a death wish. . . . In addition, around Halloween he became explosive to the point of violent with an older brother." This would have been Donald, the only other brother still living at home. "The family feels the threat of this is im-minent, every waking minute, and the mother is 'deathly afraid of the results.'"

The doctors were proceeding on the assumption—or, perhaps, a wish—that somewhere there might exist the perfect combination of medicines to help bring Peter back to some manageable baseline. In one meeting at Pueblo, Mimi said she thought that Peter did best on a combination of lithium and Prolixin, but Don said he was concerned about a tremor Peter had seemed to develop while on Prolixin. The doctors suggested trying Peter on another bipolar drug called Tegretol in addition to the lithium. Peter agreed, though to the doctors he still seemed irritable and even paranoid.

Peter told the staff psychiatrist that he wanted to write a book about his life. He said he was going to Tibet to study the martial arts, that he had been crucified and resurrected, and that he was covered with the blood of Christ. At times, he burst into song. "I'm cured. I'm well," he said in May 1986, back at Pueblo for the ninth time. "The Priest has anointed me and healed my whole body. . . . I believe that when you are anointed with oil, that is a cause for repentance and you are healed *without* medication."

Two months later, in July, he entered the room for an interview carrying a Bible, bragging about having converted several of his fellow patients to Christianity the previous evening. But he also said that he was aware that he had a disease, and that he knew the lithium was meant to keep him from getting "too hyper, from working twenty-four hours per day, like I was doing with three different jobs." Without the lithium, he said, "my blood pumps really fast."

It was at about this time that Peter sat for an assessment that seemed to shake loose something new about him, something he hadn't discussed before. He started out irreverent, as usual. Asked about his marital status, he said, "I divorced the United States." Asked if he had any special vocational training, he said, "I'm in the Federation," a nod to his father's old organization. Asked if he had any allergies, he cited both lithium and Prolixin.

Then came some boilerplate mental health questions.

Have you been hearing voices?

"Voices from God. He tells me to obey the commandments and love one another."

Have you been experiencing suicidal ideation?

"Yeah, 'cause if I get hold of a knife or a spoon, I'll swallow it. I took a whole bottle of lithium once."

Have you ever hurt anyone?

"Yeah, all kinds of people."

Have you ever been involved in physical or sexual abuse?

"Yes," Peter said. "I was abused as a child by my brother. I won't say which one."

—⁂—

The brothers who had not become sick had been doing their best to move forward with their lives, with varying degrees of success.

John, the devoted classical music student, thought Boise was completely unimpressive, the middle of nowhere, when he first moved there in the 1970s. Then he went fly-fishing for the first time, and he noticed no one was in his way. That was when he knew he'd found a new home. John was, in his way, the embodiment of the dual nature of the Galvin family: outdoorsy but scholarly; athletic and capable, but drawn to a life of the mind. He was the only Galvin brother to put those childhood piano lessons to use and earn a steady paycheck, teaching music to elementary schoolers. His trips home to Colorado with his wife, Nancy, also a music teacher, were so infrequent, John would say, because the expense of visiting was too much for a family of teachers.

But it also was more convenient not to visit. What was happening to his brothers completely terrified both John and Nancy. Every trip back to Colorado had a way of justifying those fears. Once, they left their two small children at the house for a few hours of baby-sitting and came back to see flashing lights in the driveway. There had been another blowup with the sick boys, and Mimi had taken John and Nancy's kids into a closet until the police came. The kids were fine, but John and Nancy's visits grew less frequent, and they never stayed overnight at Hidden Valley Road again.

To some of his brothers and sisters, it seemed like John had all but abandoned the family. But the truth as John saw it was that he felt distanced from them—robbed of having a family at all by the unpleasantness of the disease. When the time came for them to tell their own children that they had half a dozen mentally ill uncles—and that the family might have a genetic legacy that could affect them one day—John and Nancy said nothing. His sick brothers were never a topic of conversation at home. His son and daughter would not learn much about the family illness until they were in their twenties.

Michael, late of the Farm, had made a life for himself in Manitou Springs, the hippie-friendly town next door to Colorado Springs. He married and had two daughters, and then divorced. He made a living on this and that—helping take care of older people, home repair jobs, the occasional classical guitar gig. He remained skeptical of the medical establishment's treatment of his brothers—still raw from his own misdiagnosis years earlier, still suspicious of any conformist impulse, still thinking hopefully that his sick brothers had the power to snap out of it.

Richard, the rehabilitated teenage schemer, had more of Mimi's grandfather Kenyon in him: cocky, restless, impulsive. All that was largely a front, of course: Brian's death had made him wonder if it was only a matter of time before he, too, went crazy. "It scared the shit out of me," he said. "For twenty years, I anesthetized myself, hoping that it wouldn't happen. I blocked my family out." His teenage wedding, prompted by a pregnancy, had resulted in a very short-lived marriage. Richard worked to maintain ties with his son after the divorce, but also spent much of his twenties partying, playing jazz piano at local clubs at night, and sharing his cocaine with his younger brothers and sisters when they asked for it. He avoided home more and more, getting filled in on the latest crises when he visited for Thanksgiving or Christmas. "I was hearing these horrendous stories—'Oh my God, you won't believe what Donald just did,' or what Jimmy just did or what Matthew or Peter or Joseph just did."

In his early twenties, Richard was hired by a mining company that was connected, indirectly at least, to the Koch and Hunt oil families. He spent many years working those connections to score different gigs and round up investors for mining projects around the world. Richard had never wanted his family's issues to taint his career prospects, so he kept his distance. Only every now and then did what was happening to his brothers break through and register with him. In 1981, when Ronald Reagan was shot, Richard happened to be acquainted with a close relative of John Hinckley Jr. He heard about the FBI descending on Hinckley's family, gathering facts, asking questions. The thought entered Richard's mind before he had the chance to stop it: How soon before he got a knock on the door like that?

In the mid-1980s, Richard said, he bought a mine that became a Superfund site right after he got control of it. The litigation with the previous owner took two decades and cost Richard $3 million to resolve; then came a bankruptcy. All the while, he continued to make other deals and boast about his success to his siblings. When the genetics researchers came to examine the family, Richard did the minimum, giving some blood and sitting for an interview, but he kept his distance from the medical efforts after that. He would see his mother alone, presenting himself as a welcome distraction for Mimi—an entertaining visitor who could get her mind off of her troubles. To his surprise as much as anyone's, he found himself enjoying a warm relationship with Mimi—years after viewing her as a harsh disciplinarian the way his other brothers had.

Mark, the eighth son, had once seemed like the brightest of the boys, able to beat his older brothers at chess at the age of ten. As a child, he had been the peacemaker in the family, the one who tried to break up the fights. "I think I was kind of like Mom's little angel," he once said. "Maybe she was less hard on me than on all my brothers. I could do no wrong." But the loss of so many brothers weighed hard on Mark. He dropped out of CU Boulder, married, had three kids, and never returned to college. He divorced and remarried, and eventually found stable work as a manager of the University of Colorado bookstore. "I think what he did was he decided that he needed to take all the pressure off of himself and to lead a very simple life," Lindsay said. "That was his solution."

Mark remained in close touch with his sisters and his parents, and was given to moments of heavy sentiment, often prone to crying when thinking about the old days. He may not have caught the family illness, but it had essentially marooned him. Joe and Matt and Peter were his teammates, the ones he spent every waking moment with as a boy. They were the hockey brothers, and everyone else in the family had been little more than background players. Once they had their psychotic breaks, one after the other, it was as if the three most important people in the world to Mark had fallen off the face of the earth.

———

THE PATRIARCH OF the Galvin family was in his sixties now, but aged considerably more by his stroke, plus some more recent health concerns. In the 1980s, Don received his first cancer diagnosis: a carcinoma about the size of a nickel on the top of his head. The cancer spread, and he'd have three treatments over the next fifteen years, including a dissection of his breast to pull cancerous tissue out of forty-five lymph nodes, an operation on his prostate, and polyp removals from his colon. By the 1990s, he was on medication for hypertension; from there, congestive heart failure would be just around the corner.

Gone were the days of traveling the world for the military and defending the country with NORAD and flying to meetings and parties with politicians with the Federation of Rocky Mountain States. At home all day now, Don collected maps of Alaska and sat for hours planning expeditions to find goshawks with his old falconry friends. The plans were a fantasy. Don's ankles were too swollen, his heart too clotted, his mind too compromised for such a trip. But Don talked on the phone with those friends, and wrote them cards and letters, holding on to the idea of falconry as a shorthand for the man he'd been, envisioning himself in the ranks of the kings and ornithologists and naturalists who turned the taming of a wild bird into something sublime. Without that, he might have had nothing at all.

It wasn't enough that everything Don had built up in his life— his academic scholarship, his rank, his expertise—now seemed to mean nothing. Where he once looked at his children after a day at work and imagined that he was a part of something important, the proud leader of a storied tribe, now he could only look on in wonder at everything that had happened. Don would sift through photo albums with visitors, smiling and pointing to snapshots of one sick son or another, wryly mentioning how much each received from the government in benefits every month. "This one gets $493, but *this* one gets $696. . . ."

Lindsay sometimes wondered if her father was disappointed in all of them—feeling that even the six who had escaped mental illness had, in some way, failed him. At one time or another, John, Michael, Richard, Mark, Margaret, and Lindsay each believed they would never measure up to the man their father once was. They all felt those same expectations, and they all thought they'd fallen short.

CHAPTER 29

The Galvin sisters were together in Boulder, sitting around one night, when Lindsay finally brought it up. It was, in her view, a risk. She still remembered the other time she'd mentioned it to her sister, how she had been shut down.

This time was different. This time, Margaret said, "You, too?"

Margaret had no memory of Lindsay ever asking about Jim before. That was how determined she had been to live as if it was not happening at all. But now they were both ready. They compared the details of what Jim had done—parallel traumas, each happening without the other knowing. At first, they were amazed at how similar their experiences had been, like noticing a twin who in some way had been around the whole time.

Then they both felt a little drained—filled with dread and even regret about ever having said anything at all. Talking about it made it even more real.

In time, that, too, gave way to simple, grateful relief—just knowing that someone else knew what they were talking about and understood the depth of the pain. That they lived in the same family and understood what had happened in the same way was a rare bit of good fortune for them both. After years of first avoiding each other and then tiptoeing around each other, each sister not wanting to burst the other's bubble, Lindsay and Margaret found that they were able, finally, to offer each other comfort.

For years, the sisters talked about everything. *Do you remember this? Did this really happen? Remember that night?* They cooked together, exercised together, and deconstructed their childhoods together. This was the period when they were closest, bonded by a mission to understand what had happened to them.

They made a promise to each other that if either of them even felt remotely suicidal, she would call the other one.

Lindsay told Margaret about the help she was getting now—how she talked to Tim, who talked to Nancy Gary, who helped her find the right therapist. Margaret listened carefully. Lindsay recommended a book: *The Courage to Heal.* Margaret promised to pick it up.

MARGARET'S MARRIAGE TO Chris had lasted barely a year. They had honeymooned in Greece and then Cairo, where they had a personal guide arranged by Chris's father, the oil executive. Not long after they came home, Margaret discovered she was pregnant. She hadn't planned it, and now that it happened, she didn't know what to think. When Chris demanded that she get an abortion—and threatened to leave her and the baby if she didn't—she was well on her way to understanding that they had no business being married.

She went ahead with the procedure, and the marriage ended abruptly when Chris filed for divorce—nine months after Margaret discovered the pregnancy, an odd bit of timing that did not go unnoticed by Margaret. She moved back to Boulder, living close by her sister and trying to finish college and start over. She got her degree in 1986, but not before another lopsided relationship, this time with a mushroom-dealing climber with taut muscles, a broad back, and piercing blue eyes who would wake and bake every morning before heading off to Eldorado Canyon for the day. Sometime after graduation, Margaret shook this boyfriend loose. Good riddance, he'd told her; as far as he was concerned, both Galvin sisters were a buzzkill.

> The moon must be in Scorpio. . . . I am trying desperately
> to feel better and the desperation is killing me. My feelings
> are deadened and my reactions to situations have not been

the greatest. Maybe it's because I have not reacted enough. I don't know.

Margaret's diary, April 23, 1986

Margaret found an outpatient rehab, visiting a counselor there once a week. Instead of smoking pot each morning, she headed out for runs up Flagstaff Mountain. She found a job at a tchotchke shop on the Pearl Street Mall, started yoga, and contemplated new ways of thinking, phrases like "stepping into the softness of myself." It was, she sometimes thought, like making a new friend.

But unlike her sister, Margaret had no real desire to delve into her family issues or seek deeper therapy. She wanted to be gentle with herself. She went camping and mountain biking in Moab in Utah, captivated by the immense red rocks all around her. She rode more than a hundred miles along the White Rim trail inside Canyonlands National Park—four days and three nights. The *Nutrition Almanac* became her new bible; she did almost all of her shopping at Alfalfa's, the only health food store in Colorado at the time. Slowly she felt able to look closely at some of the things she'd been running from for years. The pain of dealing with her collapsed marriage. The residue of years of sexual abuse by Jim. The unresolved issues with her entire family.

She and Lindsay became roommates in a new place, a condominium where they split the rent. *We are really lucky to have each other,* Margaret wrote in her diary in 1987, *and we have to remember that always.* Wylie came out to Colorado to visit. He was the stable one, the one she'd known before her marriage and was a better fit for her. He was working on the trading floor at the Chicago Mercantile Exchange. He wanted to be with her. He always had.

Margaret still was afraid that being with Wylie would mean being honest about everything about herself. *He's pretty cool,* she wrote in her diary, *but he engulfs me with self-disclosure and it makes me retract.*

MARGARET CONFRONTED JIM a few years after Lindsay did. While Lindsay had done it in person, if somewhat on the fly, Margaret did it over the phone, a safe enough distance away.

Jim denied everything, just as he had with Lindsay. And when Margaret opened up to her mother about Jim, Mimi reacted the same way she had with Lindsay: She shared her own experience with her stepfather, and then gave Jim a little benefit of the doubt because he had been sick.

Margaret was so angry she could barely function for weeks. She was teetering at a great height now; she could fall off one way or the other. If she stayed fearful and embarrassed and ashamed of her family, she thought she might never make it out alive. But she was not sure of any other way.

Wylie was there for Margaret now. She needed someone she trusted to stand by her while she recalibrated what sex and intimacy meant to her. They lived in Chicago together for a few years and then they moved back to Boulder together. They married in 1993 and started a family as she continued to search for a way forward.

She found a therapist, referred to her by Lindsay's therapist, and supplemented that with countless nutritional and exercise regimens and nontraditional forms of therapy—the latter being something of a town specialty in Boulder. She practiced art therapy with an esteemed teacher from Naropa University, and meditation with a Buddhist instructor. She trained in the Hoffman Process, a retreat-based amalgam of Eastern mysticism, Gestalt, and group therapies, in which she indulged in creative visualization—turning her turmoil into a dragon, then trying to slay it. For a few years, she found solace in Brainspotting, an avant-garde trauma therapy concentrating on controlling one's eye movements in the midst of creative visualization. An offshoot of the better-known Eye Movement Desensitization and Reprocessing, or EMDR, therapy, Brainspotting is meant to help a patient relive traumatic events, only this time with a sense of control and safety. ("The child whose memory we're activating is getting nurtured," her therapist, Mary Hartnett, would say.) In sessions tracking the direction and focus of her vision, Margaret ran through the whole catalogue of traumatic memories, starting with the smaller items: Jim slashing Lindsay's car tires the night before her wedding, Donald naked on the floor of the empty house, all the furniture moved into the yard, Matt stripping naked at the Garys'. Gradually, with her therapist's guidance, she worked her way up to the major traumas: Jim's sexual abuse, Brian's murder-suicide. During sessions, Margaret would sometimes cry for an hour and a half, grieving the loss of the life she could have had if they all had been normal. Then she would go straight to bed and sleep through the night.

In rethinking her life, Margaret kept coming back to her mother: Why did Mimi have all those children? Why did she protect the sick ones at the expense of the well ones? Why did she put both of her daughters in harm's way by sending them on weekends to stay with Jim, whom she knew was insane? Slowly, she did her best to see her mother in a new way. She began to think that Mimi hadn't been capable of seeing the sexual abuse going on right under her own nose because she, Mimi, had never really acknowledged her own abuse. Could that also have been the reason Mimi kept having baby after baby after baby, with no sense of limitation, no sense of scale or proportion? Her mother had been binging on family—running away from the past and trying to build something ideal. Something flawless.

For the first time in a long time, maybe ever, Margaret felt a kinship with her mother, as a survivor. She was getting closer to healing. But with the exception of her sister, she needed to keep her distance from the family to get it done.

DON
MIMI

DONALD
JIM
JOHN
MICHAEL
RICHARD
JOE
MARK
MATT
PETER
MARGARET
LINDSAY

CHAPTER 30

Lindsay went to the jail in Boulder and took a long, careful look at her brother. Peter was thirty-one but still could pass for a Colorado college kid, with his ruddy complexion and Eddie Bauer–ish wardrobe—goose down jacket, wool socks, hiking boots. He was, in many ways, the same rebellious spirit he'd been as a boy—bright and chatty and charming and always in trouble, fighting with his parents, ping-ponging between Hidden Valley Road and the hospital. But this pattern was intensifying, and Peter seemed lost, unable to modulate himself even a little now, and Lindsay was finally in a place in her life where she thought she might be able to help him.

At home with Don and Mimi, Peter had been flying into rages, once shattering most of the windows in the house. On one trip to Penrose Hospital—where, once alcohol withdrawal set in, he felt bugs crawl on his skin and saw maggots drop from the ceiling into his mouth—he was written up for "being inappropriate sexually with nurses on the ward" and even trying to assault one of them. Out in the world, Peter lived on the streets or crashed with acquaintances; once, after being arrested for reflecting a light into the eyes of drivers as they passed him on the side of a road, he announced that he was a pilot and that he needed to rescue the city. In his more fantastical moments, he'd vow that he was going to run the Federation of Rocky Mountain States just like his dad once did—reclaiming the throne for the family.

He had come to Boulder when he decided, in the middle of a court-ordered stay at the state hospital in Pueblo, to walk out and hitchhike to visit his little sisters. He ran into trouble as soon as he got there. On May 18, 1991, Peter was spotted shoplifting a pack of cigarettes at a 7-Eleven. When someone from the store chased him out, Peter sat down in front of the entrance and refused to move. Two police officers came by, and when one asked his name and date of birth, Peter replied "1851" before making a break for it. When the officers tried to stop him, Peter panicked and threw punches, hitting both of them in the face. Peter later would say he was just trying to shake them off. But the police cuffed him and charged him with second-degree assault on a police officer.

After Lindsay visited him in jail, the court transferred Peter to Pueblo, then back to jail in September with the recommendation that he be found incompetent and the criminal case not proceed. Lindsay seized the moment: She got Peter out on bond and took him home. She had a plan. She thought that the support of a sibling, combined with some therapy, might help mainstream Peter, break him out of his home-to-Pueblo revolving door. He was the youngest Galvin brother, and so Lindsay imagined that there was more hope for him than the others, that he might not be too far gone. Caught in the institutional pipeline since he was fourteen, Peter seemed to Lindsay to be little more than a victim—both of the system and of the Galvin family.

Lindsay told her mother. She expected Mimi to be territorial, wounded, defiant, defensive. Instead, she was fearful.

"Oh, Mary, you don't want to take that on."

"I have to try," Lindsay said. "I mean, if I don't try, then I'll always wonder."

LINDSAY HAD BEEN seeing her therapist, Louise Silvern, for seven years. After college, she put her marketing degree to work at the Eldora Ski Area, helping stage events like the World Alpine Ski Championships. Within a year or two, she became the sales director of the ski area, coordinating and mounting corporate events for the resort. While working at Eldora, she met her boyfriend, Rick, whom she

Margaret, Peter, and Lindsay, on a visit to Hidden Valley Road

would marry several years later. She was ready for a deeper relationship now, able to imagine that sort of life for herself at last. And in
1990, with just three years of experience, Lindsay used her contacts
from the ski resort to go out on her own, founding a corporate event-
planning business. The greatest moment of her twenties might have
been when she took Don and Mimi out to lunch and surreptitiously
picked up the tab.

She could work all the time, her energy not flagging. But she never
felt successful for long. There was always something a little wrong,
something that needed tweaking. Finally, in sessions with her therapist, she realized how guilty she felt about the cruel accident of fate
that had spared her, but struck her brothers. She wanted to right the
wrong, even the score. But above all, she was so high on what therapy
was doing for her that she could not help but wonder if the best thing
for all of her brothers would be to stop sending them to Pueblo—and
offer them the same sort of help that had changed her life.

Lindsay's new idea—keep the boys out of Pueblo—was intertwined with mammoth criticisms of her mother. Lindsay was convinced that throughout her childhood, Mimi had used the idea of
Pueblo as a cudgel to keep the older boys under control—to hobble

them, infantilize them, keep them prisoners of their own worst failings. But most of all, when it came to her brothers' treatment, Lindsay believed, much like Michael did, that there was a piece of the puzzle missing, and that what her therapist had given her—the chance to tell her own story and to recover—had never been made available to the boys on Hidden Valley Road.

She believed that her brothers had been ill-served, even penalized, by a profession that seemed to do nothing but give them drugs. However well intentioned prescribing neuroleptics might be, the drugs seemed to Lindsay to be little more than another kind of lobotomy—a way of warehousing people's souls. What if there was another way? What if someone asked her brothers what they thought they needed, and really listened to the answer?

More broadly, Lindsay wanted a new role in her family. Now that the need to avoid Jim was no longer keeping her away, she believed the next step in her recovery, in the process of taking control of her life, would be to try to go back and help those who seemed left behind.

Why wasn't it me? Lindsay would think. *I owe him something because it wasn't me.*

MARGARET WAS IN no mood to help Lindsay with her new project. She said the idea seemed too ad hoc to her—a fly-by-the-seat-of-your-pants thing. She had a point.

But the truth was, Margaret also was afraid. She was closer in age to Peter than Lindsay was. As a child, she had been the target of a lot of the teasing and taunting and bullying by the four hockey brothers, Peter included. Now, Peter was smashing windows and hitting cops? The very thought of being around him made her feel exposed. Her reaction was the complete opposite of Lindsay's: *I don't want it in my life. I can't have it in my life. I don't want to be anywhere near it.*

Lindsay couldn't understand that reaction. She wanted to be in the eye of the hurricane, even if part of her suffered because of it. She was dying for a chance to confront her parents, to show them how wrong they'd been, how poorly they'd handled everything—to be the master of her fate in a way she had never been as a child.

Where Margaret wanted to reconstruct the sense of a normal childhood, to get back what she lost, Lindsay had resolved never to be a child again.

———

LINDSAY BECAME PETER'S state-designated caretaker—a fair amount of paperwork for a twenty-six-year-old with a full-time job. But now she could manage Peter's benefit checks, arrange his therapy, and apply for federal Section 8 housing. She worked for months with the Boulder sheriff's office to clear up Peter's criminal charges. He returned to Pueblo in December 1991 for six months, then came back to Boulder with the plan to stay in her care.

Lindsay took Peter to all of his appointments, starting him in therapy at CU Boulder, where they didn't charge him, and introducing him to the Boulder Mental Health Center, which believed in complementing medication with therapy. They attended sessions together, and she saw how Peter seemed pleased to have a place where his feelings were acknowledged, where he was made to feel worthwhile and deserving of sympathy because of what he had endured.

She learned more about her brother's condition. While most schizophrenia or bipolar patients eventually give up and give in to the system, Peter never stopped fighting it. That much wasn't a surprise to her. But what did interest her was the reason. Peter's doctors said that while any number of patients lash out at being compelled to do anything, Peter was among the rare ones who took aim at the systemic issues at the heart of his troubles, speaking out against a medical structure that he believed was keeping him from getting better, or at least from doing what he felt entitled to do. As a result, he resisted even more—and ended up sicker, perhaps, than he otherwise would be.

Learning this made Lindsay more sure than ever that she was doing the right thing by taking in Peter. By caring for him in Boulder, she could break the cycle of resistance and illness, and help him regain a sense of control over his own life. She and Peter had a shared mission now—a way for both of them, together, to say, *Our family matters. Don't sweep us under the rug. Why can't life be different for our family?*

LINDSAY'S BOYFRIEND, RICK, took Peter skiing and ice-skating, and Peter's muscle memory kicked in and he relaxed into himself, his body suddenly reconnecting with the years he spent playing hockey with his brothers. "It was like a different guy," Rick remembered. "His tone of voice, his confidence. His mojo was on the ice." These were happy, encouraging moments, when Peter seemed willing and able to reach back and reclaim part of the person he'd once been.

It seemed to Lindsay that Peter wanted desperately to prove to everyone, including himself, how well he could be—how he was not lost. And Lindsay thought she saw improvement. Peter was a loudmouth, impulsive and overbearing, but he was also high-spirited and charming. He wasn't delusional, as a rule. He knew reality. He could hold a simple job.

One of Peter's case managers in Boulder observed that Peter seemed dedicated to becoming part of the solution—aware of the mental health care system around him, and its shortcomings, and dedicated to helping improve that system. Lindsay brought him to a meeting of CAMI, the Colorado Alliance for the Mentally Ill, where Peter spoke movingly about his brushes with the police, and the need for special training to be sensitive to people like him, to not seem threatening, to not provoke.

Lindsay believed that Peter saw how she and Margaret had made it through their childhoods alive and well, and he started to think he could, too.

Good stretches like this would last for a while—a month, maybe more—until Peter became so confident that he would stop taking his prescriptions. Then he would stay up all night, speaking quickly, hardly pausing for breath, spinning the same old fantasies about how he was going to run Dad's Federation. He would ride his bike to the top of Boulder Canyon and back, then again, and again, and again. Still anxious, he'd turn to booze or pot or something stronger to self-medicate. Then he'd spend all day on the Pearl Street Mall, the main pedestrian drag in Boulder, sitting with the street people and playing the recorder, and quite often bringing his new friends back to Lindsay's apartment to party.

That was when law enforcement would get back involved in his case. Instead of Pueblo, he'd go to a state hospital in Denver called Fort Logan, until once again he was well enough for Lindsay to bring him home.

ONE NIGHT ON Pearl Street, Peter looked up from playing his music and saw a little boy watching him. Next to the boy was a man he recognized. Peter smiled.

"Hi, Dr. Freedman!"

Robert Freedman knew the family well by now, having tested most of the siblings' sensory gating skills at his lab in Denver. But he hadn't known that Peter was in Boulder. Now, when Peter ended up in Fort Logan, Freedman would make a point of treating him and debriefing Lindsay on how her brother was doing. After several visits, Lindsay started hearing Freedman use the term *brittle* to describe her brother. That meant that the smallest little thing—a bad night's sleep, skipping one dose—could cause another psychotic break.

Freedman told her that this was the result of years of noncompliance—not just refusing to take medication, but being prescribed the wrong medication when he'd been diagnosed first with schizophrenia, then schizoaffective disorder, and finally bipolar disorder. The entire concept of noncompliance seemed to blame patients, but what especially pained Lindsay was the sense that she might have been too late to help her brother—that for years, Peter might not even have been getting the right medicine. If, in fact, there was a right medicine at all.

Even worse, when she looked at her other brothers, Lindsay saw how years on supposedly the right medicines were making them brittle, too—frailer, more withdrawn, less able to handle the slightest variation in routine. She came away thinking all of her brothers were damned if they did and damned if they didn't.

LINDSAY'S EXPERIMENT WAS starting to look like a failure. Nothing she did steered Peter clear of the revolving door for long. Freedman warned her that her brother would continue to get a lot worse over time, and that the better doctors for him were not in Fort Logan, but back at Pueblo.

From Freedman, Lindsay learned that some researchers believed that a genetic predisposition toward schizophrenia—a vulnerability, as articulated by Daniel Weinberger's developmental hypothesis—could be triggered by an environmental stressor. Maybe there was nothing Lindsay could have done to help Peter deal with his particular stressor, whatever it might have been.

But when she thought about that mixture of nature and nurture, Lindsay decided that, assuming she had the same genetic vulnerability as her brothers, she was living proof that the environment matters:

After experiencing her own trauma, she got the proper treatment, and she never got sick the way that they did. Her trauma was sexual abuse, but her brothers each had their own: Donald when his wife left him, Brian when he and his girlfriend broke up, Joe when his fiancée left him, Matt after two significant head injuries (one from hockey, the other from the time his head smashed into the patio during a fight with Joe).

Peter's trauma had seemed easy enough to spot: At the age of fourteen he had watched his father have a stroke; his first hospitalization had been a matter of weeks after that. But there was something else. Since they were closer now, Lindsay asked Peter if, like she and Margaret, he'd ever been sexually abused by Jim. Peter said yes, though he did not elaborate.

Lindsay was not exactly surprised. It seemed as if Jim had taken liberties with every young child around him. But wasn't this trauma her trauma, too? Just like that, after years of effort, Lindsay was back to wondering what it was about her—her brain chemistry, her genes, her deep dive into therapy—that kept her from ending up just like Peter.

DON
MIMI

DONALD
JIM
JOHN
MICHAEL
RICHARD
JOE
MARK
MATT
PETER
MARGARET
LINDSAY

CHAPTER 31

It never stopped amazing both Lindsay and Margaret how to so many people outside of Hidden Valley Road, their mother, in her advancing years, seemed almost saintly in her devotion to her family. "Despite some physical illness on her part, she does not seem to let this get her down," one Pueblo doctor wrote in 1987. "Her attitude is that she must keep going and somehow things take care of themselves."

On visits with doctors at Pueblo or at the outpatient Pikes Peak Mental Health Center, or Penrose Hospital, or the CARES facility where her sons sometimes stayed, Mimi never failed to impress, entertaining the doctors with stories about the opera and Georgia O'Keeffe and her grandfather and Pancho Villa. "She was always very pleasant," remembered Honie B. Crandall, a psychiatrist who, as the medical director at Pikes Peak, treated nearly all the Galvin brothers at one time or another. "Never saw her out of control, or unpleasant. But she was always saying, 'You've got to drop everything and come do this now. Come take care of this.'" Mimi was a happy warrior again. Only the war had changed.

Alone with her sick sons, Mimi's fuse was a little shorter than outsiders might have thought. She'd snap at Matt's poor hygiene, and fume about Peter's insolence, and pick on Joe for putting on so much weight. She had slightly more patience for Donald, still the son with whom she had the closest contact. After many years of trying to live in a group home, Donald had given up and come back to Hidden Valley Road, seemingly for good. "He just couldn't tolerate being with other ill people," Mimi would explain—not her exceptional son. Donald's hands had a tremor now; the doctors diagnosed him with tardive dyskinesia, a common side effect of antipsychotic drugs, causing involuntary stiff, jerky movements. Donald's explanation for the tremor was that he got it because his father "made us stand at attention because he wanted us to be doctors."

So much of what Donald said on any given day still was not linked in any understandable way to reality. But with the benefit of the same medications that were slowing him down, Donald had periods of lucidity. On a good day, he and Mimi would go bird-watching, and Donald would get slightly more animated when he saw something— "Oh, there's a red-tail!" or "There's an eagle!"—and he would reminisce about flying falcons with his father. Mimi would take him on every visit to see other relatives—her escort for the day, usually sitting quietly by her side until it was time to go. And yet as the years went on, Mimi started to grow weary of Donald's more obstreperous side. She had to hide the family photo albums to keep Donald from pulling out pages and destroying them. He smashed a large statue of Saint Joseph that had been at her fireplace for years. On one outing with Mimi to the bank, he told a teller he wanted to open an account and change his name. But on most days, Donald would not leave his room. Even at Christmas, he would greet everybody with a hug and then retreat to a hiding place. One of Mimi's granddaughters found him once, when she was about five: "Mimi"—many of the grandchildren, adorably, called her by her first name, following the lead of Margaret and Lindsay—"Donald is sitting in the closet."

Even in those moments, Mimi's heart went out to Donald. "Holidays are extremely hard," she said. "Everybody's getting together and discussing where they're going and what they're doing and how many children they're having, and so forth. It's a very hard time." Hard for her, too, to be reminded of everything she'd once hoped for him. When she looked at Donald, Mimi often would reference the boy he had been before he was sick. "People would say, 'Oh, he has such beautiful manners.' Little did they know."

In conversation, Mimi started citing a book she'd been given called *Saints, Scholars, and Schizophrenics,* about communities in Ireland where the mentally ill are cared for, and even treated as people with special insight into an otherwise unnoticed world. Just knowing such a place existed was a comfort to her—the suggestion that there might be something distinctive about Donald and the others, to make up, in some small way, for what was lost.

When the illness first took hold of the family, Mimi's life had changed, too. It was as if an entire future she'd once counted on like the sun coming up in the morning just never came to pass. She never complained about any of that directly. But sometimes when her two daughters visited, they noticed a new bitterness to Mimi. The stories she told changed: Her monologues weren't just about Howard Hughes and Jacques d'Amboise. They were about how she had wanted her husband, their father, to be a lawyer, but he insisted on the military; how she had always wanted to live on the East Coast, but Don took her around the country and out to Colorado; how she never thought that she would have twelve children, but Don wanted twelve, and so they had twelve. She did what a wife does, she said, even converting to Catholicism, because that was her role. She had served everyone else, she said, enumerating the great sacrifices she had made to do so.

At her worst, she blamed Don's side of the family for the illness. The son of one of Don's brothers seemed to be unbalanced now, perhaps bipolar. It was a matter of time, she would say, before science proved that what happened to her children was an inherited illness from the Galvin side.

This struck both Lindsay and Margaret as petty, even cruel. Their father was a shadow of the man he'd been, spending most of his time in front of the television. When the subject of any of the sick boys came up, he seemed unable to look closely at the situation anymore. And his chin would quiver when anyone talked about what Jim had done to the girls. He stopped short, at least in Margaret's estimation, of taking responsibility. But he was no longer quite as distant as he once was. He teared up. Why pile on him now?

Something was bothering Mimi—and it had to do with her daughters. She knew that in their eyes, she was both the villain and hero of the family: a mother in denial, heartlessly neglecting her daughters because she was so attached to her sick sons, and a mother who kept her family together, left to care for so many sick sons by herself. Mimi sensed she was being judged. It frustrated Mimi that as much as she took on, the last people to appreciate her seemed to be some of her own children. She could only abide this for so long.

IN THE 1990s, a revelation hit Mimi that she never saw coming—something devastating that, the more she thought about it, made dreadful sense. Seemingly out of nowhere, Donald confided in his mother that, as a teenager, he had been a victim of sexual abuse. And when Mimi asked the name of his abuser, the answer was a man whom she had considered a close friend.

In the late 1950s, when he was just a boy, Donald had been the first of the Galvin sons to serve as an altar boy at St. Mary's for Father Robert Freudenstein—the same priest who had instructed Mimi in Catholicism and baptized her. In the years that Freudy became close to the family, a confidant to both Mimi and Don, young Donald was close to him, too. When Donald was sixteen, he had stayed out on the prairie with Freudy for a week, chauffeuring the priest in his car after he'd lost his license. Now Donald was saying he'd been molested by him.

Mimi had no idea how to react. She was almost seventy now; how many more horrors was she supposed to bear? And Donald always said so much, almost all of it nonsense. She tried to ignore it. But Donald continued to insist, in his flat, deadpan way, that it was true. And the crisis of sexual abuse in the Catholic Church was all over the news now. From the publicized cases it seemed that most people were coming forward decades later, just as Donald was, having been silenced by shame and in some cases intimidation.

Father Freudenstein had never made the news in this way. But Mimi could not stop thinking about it. To think that this happened to her son while she was supposed to be protecting him brought her lower than she'd been in years—since, perhaps, the death of her son Brian. The more she thought about Freudy, the more she saw how invasive he'd been, how he'd made himself indispensable to her, how she came to trust him to be alone not just with Donald but with all of her older boys. And the more she learned about priests and young boys, the more Mimi began to wonder how many of her sons might have been victimized.

At first, it seemed like there was nothing to be done. So much time had passed, and Donald was Donald—diagnosed with schizophrenia, heavily medicated for decades. But Donald repeated what he'd said to anyone who asked. He never wavered. The other brothers had varying memories of Freudy. While John remembered being teased by him, Michael and Richard recalled liking him. Richard remembered Freudy taking his older brothers—Donald, Jim, John, and Brian—hiking up in Glenwood Springs for two days at a time. "Mom and Dad were relieved," Richard said. "They had a trusted priest."

It was Richard who, entirely by happenstance, learned more about Freudy. A close relative of Richard's girlfriend Renée, a man named Kent Schnurbusch, told the couple that he had known the priest as a teenage boy in 1966; he'd been groomed by Freudenstein, he said, and had sex with him. Years later, Kent attended a meeting of the Colorado chapter of the Survivors Network of those Abused by Priests (SNAP) and mentioned Freudenstein's name. Two different men said they had heard of Freudy; he was gay, they said, and suffered from alcoholism, which alone might have explained why he was transferred so often to small parishes and never rose up in the ranks of the Church. Freudy had retired from the priesthood in 1987 and spent his final years in severe decline before his death in 1994.

Kent decided to go to the chancellery and make his claim, to see what else there might be to learn about the priest who had taken advantage of him. The meeting was so brief, it took his breath away. Instead of pushing back against Kent, the priests at the chancellery simply asked him how much he was expecting in damages. He was unprepared for this. He wasn't there for the money so much as the closure. He asked for $8,000, and the chancellery gave him $10,000.

When Kent told all this to Richard and Renée, he was as astonished as they were that the priest had known all the Galvin boys so well, just a few years before his experience with him. Kent had been eighteen when he knew Freudy—a teenager, like Donald had been when he went out to stay on the prairie as his chauffeur.

When Mimi learned Kent's story, what was once a possibility became, to her, a certainty. Here was corroboration, and even signs of a modus operandi. It didn't matter to her that Freudy's name did not turn up on any of the lists made public by the abuse survivor and advocacy groups, or that he was never named in any public lawsuit.

Everything lined up, as far as she was concerned. Who knew what incidents weren't public, and which disgraced priests had their sins swept under the rug? Mimi came to believe that Freudenstein had been perusing her boys like boxes of cereal at the supermarket until he found the one he liked the best. "He had culled my family," she said. "He knew it was a big family of boys."

From there, Mimi seized on Father Freudenstein as a new global explanation for everything—the big reason that things went so wrong in her family. Didn't it make sense, she'd say, that the priest sexually abused Donald, who in turn physically abused his brothers, at least one of whom, Jim, went on to sexually abuse their sisters? What if Jim, too, had been molested by Father Freudenstein? Wouldn't that explain why he became a pedophile? Maybe all the schizophrenia in the family—which Mimi had, up until now, believed in her heart had to be genetic—was set into motion by the stress of this chain of abuse? Look at how Donald and Peter both became so hyper-religious in the thick of their illnesses; could that really have been a coincidence, or was the Catholic imagery in the air, ready to be repurposed in the wake of trauma?

Mimi was leaping to several conclusions, of course. Sexual abuse does not cause schizophrenia; that much is certain. Even a torrent of sexual abuse like what Mimi had envisioned still could not answer the bigger question of why there had been so much mental illness in their family. Lindsay understood how Mimi was conflating the two things, the sexual abuse and the mental illness, and she thought she knew why. Blaming Father Freudenstein had, at least for Mimi, the virtue of taking some of the blame away from her—as long as you didn't linger too long on the question of how often a mother and father would have to be looking away for an ill-intentioned priest to have so much unfettered access to their boys.

Mimi renounced her faith. She told her children she did not want a Catholic burial, and that she wanted to be cremated. She was turning her back on all that now. Time was running out. She wanted the world to know who was responsible.

SOMETIME AFTER DONALD told her about Father Freudenstein, Mimi decided, in between her usual refrain that they'd been the perfect family before mental illness struck, to become more open about the past, sharing information with her daughters that she'd never before dreamed of discussing. What neither daughter expected was that these disclosures would be about their father.

Mimi began by going into detail about episodes throughout her marriage that, she believed, offered a different perspective on Don. The first happened in 1955, she said, shortly after the family's transfer from Colorado Springs to Canada, when Don ended up at Walter Reed Hospital in Washington, D.C., with what Mimi now was saying was a deep, powerful depression. He also had a milder episode later, she said, while they were living in northern California—something like a panic attack. Don had spent so many recent years at home, becoming more and more despondent after a cascade of health problems—they'd all seen that. Now Mimi was saying she believed that Don had a history of clinical depression, through his entire life.

Neither Lindsay nor Margaret believed her, at least at first. This seemed like another one of Mimi's deflections, a smoke screen of denial to keep criticism away from her—and perhaps to even back-handedly blame the boys' mental illness on Don's genes. But without wanting to, the sisters began to think of their father differently. What if post-traumatic stress disorder from the war had seeped into everything their father did during their childhoods? Did he somehow pass along his own traumas to the boys? And the most troubling question of all: Could Don have been the source of the violent streak in the family that culminated in what Donald did to Jean, and Brian to Noni—and Jim to them? Both Margaret and Lindsay had spent so many years focusing on their mother and all that she did and didn't do. Here were a new set of questions they had never thought to ask.

The sisters were even less prepared for their mother's next announcement. Mimi said that in the years before his stroke, there had been many other women in Don's life—at least six, by her count. The first had been in Norfolk, Virginia, just after the war, when Don was traveling up and down the Atlantic on the USS *Juneau*. Mimi told both Margaret and Lindsay about how she was supposed to have gone on one of those voyages, too, with Donald and Jim, who were still little. This, she said, was the trip where Don met the wife of a senior officer, and started an affair. If Mimi had been able to take that trip, she told Margaret, that affair might never have happened. Mimi found out about it later, she said, and they transferred away from Norfolk. But Don would not be held down forever.

This surprised both sisters. But in some strange way, this new view of their father also filled a gap in their understanding of their parents' relationship. Lots of what they'd seen at home made more sense to them now. Like how their father, at the height of his powers, always seemed to be somewhere else. And those dinner parties at the Crocketts', where the neighbors' wives called their father Romeo. The more they thought about it, the more the affairs explained so much of their childhood—even, perhaps, Mimi's quest for a perfect household.

Mimi had come forward with all this now to show her daughters that Don was human, not perfect, deserving of the same scrutiny as herself or anyone else. Now it was Mimi they wanted to understand better. Why did she stay with Don all that time? Did she stay because she wanted to—or because, after she'd had the children he'd wanted, she had no choice? Why did she agree to be at her husband's mercy, while he was at liberty to do as he pleased?

Margaret thought of a painting of her mother's, now in Lindsay's possession, of Pinocchio, hanging on a string being held in the hooked beak of a falcon. For Margaret, that painting was a fair metaphor for her mother's true feelings—made to care for twelve children, while her husband was off somewhere else. She wondered if all of those traits she'd ascribed to her mother—the inability to be truly present or vulnerable—were really more her father's. Say what you wanted about Mimi, but she never left. She never stopped trying.

CHAPTER 32

1998
University of Colorado Medical Center,
Denver, Colorado

Throughout the 1990s, most of the Colorado-based members of the Galvin family—Mimi and Don, Lindsay, Margaret, Richard, Michael, Mark, and the sick brothers Donald, Joe, Matt, and Peter—went to Denver and submitted to long days of testing in Robert Freedman's lab. Whenever Freedman had the chance to discuss his research, his description of sensory gating and vulnerability, of schizophrenia brains having difficulty pruning information, made sense, at least to Lindsay. She thought of how sometimes one of her brothers would be especially sensitive to something she thought was background noise, like the hum of a fan.

Freedman had never thought of his brain-electrophysiology experiments—the double-click test that measured a patient's sensory gating abilities—as a foolproof test for schizophrenia. He saw them as one of many potential strategies for having a look inside the brains of his test subjects. With the Galvins, Freedman found that many family members could not inhibit the second click, including some non-mentally ill family members like Lindsay, but some of them could.

The next step was to see if the ones who failed shared a certain genetic trait that others did not.

This put Freedman in unfamiliar territory. He was a central nervous system guy, not a geneticist like Lynn DeLisi. "I was late to genetics," he said. "Lynn was way ahead of me."

What he did know about was brain function. He understood how the hippocampus—that seahorse-shaped swath of brain matter located in both the left and right lobes of the brain—is the part of the brain that helps with situational awareness, figuring out at any given moment where you are, why you're there, and how you got there. He'd seen, and his double-click tests had affirmed, how that process requires not just neurons, or brain cells, to bring in sensory information, but the inhibitory interneurons that erase the brain's whiteboard of situational information instantaneously. Without the inhibitory interneurons, we would end up processing the same information all over again—wasting time and effort, grinding our gears, becoming disoriented and, perhaps, anxious and paranoid and even delusional.

Now, Freedman wondered if there was something at the cellular level that these inhibitory interneurons turned on and off—a mechanism that wasn't working properly in the brothers who got sick. A section of Freedman's lab began testing the brain cells of rats and learned that the on-off circuit for the inhibitory neuron was controlled by a crucial element of a cell in the hippocampus called the $\alpha7$ (or alpha-7) nicotinic receptor. The name is complicated, but its reason for being is more or less straightforward. The $\alpha7$ receptor is a master communicator, sending messages from neuron to neuron so that the circuit can work properly. But in order to do its job, this receptor needs a compound called acetylcholine, which behaves as a neurotransmitter. Freedman wondered if people with schizophrenia had faulty $\alpha7$ receptors, or simply lacked enough acetylcholine to get those receptors to work the way they should. If Freedman was right, this meant that for some of the Galvin brothers, the machine that was supposed to keep them from losing their minds might, essentially, be out of gas.

To prove this, Freedman needed to move from rats to humans. And so in the late 1990s, he embarked on one of the first genetic studies of his career. He collected data on nine families including the Galvins—104 people in total, including 36 schizophrenia patients. Among those family members who responded badly to the double-click test, Freedman searched for a common genetic pattern. From analyzing those tissue samples, Freedman was able to track down the precise location where the receptor problem took place—a chromosome that was home to a gene called CHRNA7, which the body uses to make the α7 receptor.

In 1997, Freedman identified CHRNA7 as the first gene ever to be definitively associated with schizophrenia. He and his colleagues had made history, and more importantly, he was one crucial step closer to learning how schizophrenia functioned. Now he had to find out what was going wrong with that gene. He already had an important clue: The brains of the families he was studying, including the Galvins, had about half of the number of α7 receptors that typical brains had. The receptors they did have were working just fine. The problem was they lacked enough acetylcholine to get the switch turned on to make more receptors just like them.

MARGARET REMEMBERED CHAMPAGNE being popped as she walked into Freedman's lab. She and Wylie were there to get advice about whether Margaret should have children. Freedman and his team had just made their CHRNA7 discovery, and the doctor was happy to take a break from the celebration to explain what this new information might mean to the Galvins.

The last thing Freedman wanted was to discourage Margaret and Wylie from having children. While the brothers and sisters of people with schizophrenia do have a much higher than normal likelihood of having schizophrenia themselves—ten times the chance, actually— the same, he noted, is not true of parents and their children, or uncles and their nieces and nephews. The genetic explosion in Margaret's family, he maintained, was not necessarily an indicator of some supergene that would affect successive generations. Schizophrenia has a way of disappearing in families and then reappearing, and there was no reason to believe Margaret's children were fated to become mentally ill.

It seemed hard to imagine that their risk was as low as anyone else's, but that was exactly what Freedman was saying. But what about everything his lab had just uncovered about the gene related to schizophrenia? Freedman filled a big whiteboard with information about the place on the chromosome where Margaret's family's data had helped point to a trouble spot. Nothing about that genetic irregularity could be used to predict schizophrenia, he said. All it could do was offer a road map to what needed to be treated once it appeared. And he had a pretty good idea of how to do it.

FREEDMAN'S DISCOVERY WAS not happening in a vacuum. Dozens of other researchers were conducting other studies of mutations of other genes in other chromosomes. By the year 2000, at least five more trouble areas would be isolated, with many more still to come.

The $\alpha7$ receptor, however, stood out from the crowd because of its special relationship with nicotine. No one experiences this more vividly than habitual smokers: Nicotine has a way of turbocharging the effects of the acetylcholine that this receptor needs in order to function, and smokers—or the $\alpha7$ receptors in their brains—like it when their acetylcholine is turbocharged. This is the feeling cigarettes can give smokers—that way nicotine has of focusing their minds for short periods, or calming them. Could it just be a coincidence, Freedman wondered, that many schizophrenia patients—Peter Galvin among them—can't get enough cigarettes? For very brief moments, nicotine may offer them at least some relief from their delusions. If Freedman could amplify that effect—mimic it in a lab, bottle it, and send it out to everyone diagnosed with schizophrenia—could it treat the symptoms of the illness more effectively and less harmfully than Thorazine?

First, he needed more proof. In 1997, Freedman devised an experiment: He gave nicotine to people with schizophrenia, usually many pieces of Nicorette chewing gum, and then measured their brain waves with his double-click test. Sure enough, people with schizophrenia who chewed three pieces of Nicorette passed the test with flying colors. They responded to the first sound and didn't respond to the second, just like people without schizophrenia. The effects didn't last after the nicotine wore off, but Freedman still was stunned.

His study won applause from a lot of his colleagues, including Richard Wyatt, Lynn DeLisi's old boss at NIMH, who called the Nicorette experiment "important and exciting" and the promise of nicotine "intuitively very strong." Freedman went all-in on nicotine. He made plans to develop a drug that did what nicotine did to the $\alpha7$ receptor, only better—so well that schizophrenia patients could find relief from their delusions not for minutes but hours, or even days. He secured funding for a drug trial from NARSAD, the National Association for Research on Schizophrenia and Depression (now known as the Brain and Behavior Research Foundation), a donor-supported group that serves as the American Cancer Society for mental illness. "We thought perhaps we could make a better nicotine," he said.

He found a natural substance called anabaseine that mimicked the function of nicotine. A researcher in Florida had been cultivating a synthetic version with no solid idea of what use the drug might have. He told Freedman he'd been waiting for ten years for someone like him to call. Freedman cultivated the drug, called DMXBA (short for 3-2,4 dimethoxybenzylidene anabaseine), and started testing it. The drug had the same effects as nicotine in the double-click test. And when, in 2004, he tested the drug on a group of schizophrenia patients in a double-blind controlled study, the results seemed miraculous. One subject who got the real drug, not the placebo, told Freedman that she had been having trouble finishing a short story she'd been writing, but now she was able to concentrate enough to do it. Another said, "I'm not noticing my voices." The mother of a third told Freedman that for the first time, her son was able to take in the scenery around him—to be amused by watching rabbits in the yard, undistracted by his own hallucinations.

Within a year several different pharmaceutical companies were hard at work creating versions of his drug. They couldn't just buy his because its patent, owned by the University of Florida, had been in existence too long: No company wanted to buy a patent that was just a few years away from expiring. "There isn't much financial incentive for using the drug that we've gotten to work in clinical trials," Freedman said, "so they had to go out and make their own."

As an unpaid advisor, Freedman told each company the properties of the drug, hoping they would design their versions using the principles he suggested. A few companies made it pretty far. One company, Forum Pharmaceuticals, worked on trials that were halted after too many subjects experienced constipation. Another company, AbbVie, the research division of Abbott Laboratories, made it to the third phase of clinical trials with a drug based on DMXBA with mixed results, and then stopped their research. The problem, as Freedman saw it, was that they insisted on a once-a-day dose. Freedman's team had tried that, but found that his drug only worked when administered in three or four small doses over the course of a day. Abbott thought it would never be able to market a drug that had to be taken that frequently, on such a rigorous schedule. (Think of Peter Galvin, skipping out on his drug regimen constantly, only to have yet another psychotic break.) Their once-a-day dose failed, too. "I think the company's pharmacologists were smart enough to know all this," Freedman said, "but their marketing people rule how they make drugs. And so, they sort of were doomed to failure."

Freedman saw the experience as an object lesson in how pharmaceutical companies work. "It was disappointing, because I think they could have gotten a good drug out of what they were doing." After all that promise, he was back where he started. To jump-start the $\alpha7$ receptor and strengthen the brain's ability to process information, Freedman would have to find another way.

It had been a decade since Lynn DeLisi first met the Galvin family, and she was still steadfastly collecting families, still acquiring DNA, in hopes of finding a genetic abnormality that helped explain schizophrenia. She wasn't having much luck, and neither was anyone else. In 1994, *The New England Journal of Medicine* published a survey of schizophrenia research that concluded that very little had been learned about schizophrenia and no headway had been made in its treatment. All the doctors could do, it seemed, was what they'd been doing for years: prescribe medication and hope for the best. This, for a disease that the editor of *Nature* had a few years earlier called "arguably the worst disease affecting mankind, even AIDS not excepted."

In 1995, however, DeLisi's work attracted the attention of a well-funded investor: Sequana Therapeutics, a privately held pharmaceutical company that would eventually partner with Parke-Davis to develop schizophrenia drugs. Sequana's director of genetics, Jay Lichter, was clear about what DeLisi had to offer: "Dr. DeLisi and her collaborators have assembled one of the largest collections of families whose members include one or more sibling pairs with schizophrenia," he said. Sequana thought DeLisi had an inside track on finding a genetic link to the illness—the advance everyone was waiting for. In return, the company offered DeLisi access to the most sophisticated genetic-analysis equipment available—technology "beyond the practical capabilities of a small laboratory," she said. "As a result, we expect to move much more quickly."

With DeLisi in charge, the company funded the largest single-investigator multiplex family study to date, studying the linkages of about 350 different markers spread throughout the genome. The Galvin family's DNA was part of that study. DeLisi seemed poised for a breakthrough. But within a few years, she, like Robert Freedman, learned the hard way about the vagaries of the marketplace. In 2000, Parke-Davis was bought by Pfizer. Almost right away, DeLisi learned that Pfizer was canceling DeLisi's project. All work would stop immediately. And all the genetic material she had accumulated at Parke-Davis, including the Galvin family's DNA, would remain the property of Pfizer—unavailable for DeLisi to use, unless she found another company willing to fund the project.

Why was Pfizer not interested in DeLisi's family research? She had been making slow progress, that was true. But in research, you only have to go fast if someone else is outrunning you.

THE HUMAN GENOME Project was a highly publicized effort to map out and understand the structure, organization, and function of every single human gene—the entire DNA blueprint for building a human. The project started in the 1980s at the U.S. Department of Energy, which engaged in a sort of friendly competition with NIH to raise money for the effort. In 1990, the project launched in earnest with an estimated $3 billion in funding. This was like a moon shot for biology. If the project could successfully diagram the human genome, nothing about the study of virtually any genetic disease would be the same—even complex diseases like schizophrenia.

Before the Human Genome Project, Lynn DeLisi and others had been working with the understanding that if you wanted to go looking for the genetic mutations for schizophrenia, the easiest place to find them was in families like the Galvins. That their linkage studies had proved to be unfruitful so far, they thought, was evidence of how complicated the illness was. The alternative—searching for schizophrenia mutations by studying the genetic code of the general population—had seemed ludicrous. All that changed, however, with the Human Genome Project.

Human beings have more than twenty thousand genes that, by encoding the proteins that build our bodies and keep them functioning, play a crucial role in making us who we are—quite a massive haystack to go searching for needles in. But in theory, once the Human Genome Project collected and mapped out the genetic information of enough people, that haystack would suddenly become much easier to search. Now, all one would have to do is compare the genomes of a sampling of sick people—for any genetic disease, take your pick—with a control group, and whatever abnormality existed in the genome of the sick people would be impossible not to notice. Just like that, drug companies would have a gene to target—a specific genetic process to manipulate with medication.

With the Human Genome Project, new treatments and cures for any number of diseases seemed to be a few short years away. In 1995, the cancer researcher Harold Varmus, the director of the National Institutes of Health, organized a two-day workshop on schizophrenia at the National Academy of Sciences. Varmus, who with J. Michael Bishop had won a Nobel Prize for identifying the cellular origin of certain cancerous genes, had invited many of the usual suspects— E. Fuller Torrey, Irving Gottesman, Daniel Weinberger, Yale's Patricia Goldman-Rakic—to present their latest research. Varmus was not impressed. At some point, Weinberger recalled Zach Hall, Varmus's newly installed chief of NIH's neurological disorders division, standing up and saying, "You people have been studying this disease for thirty years, and from where I sit, you have accomplished virtually nothing."

Many of the researchers at the workshop were appalled. Some pushed back a little. Then Varmus himself weighed in, saying what he might have been planning to say all along. "You people don't get it." Everyone in the room could forget all about their enzyme research, their MRI studies, their CT scans, their PET scans. If you're not studying genes, Varmus said, "you are going to be dinosaurs."

COLLABORATING AROUND THE world, the scientists with the Human Genome Project thought the job would take fifteen years.

They finished ahead of schedule, in 2003. Not only was the recipe book for human life now readable, beginning to end, for the very first time, but along the way scientists had identified new genetic markers spaced throughout the genome that could be used for research. Whereas DeLisi had previously been limited to about a few hundred different markers spread throughout the genome, the Human Genome Project opened the door to the discovery of literally millions more. With this wealth of new markers, researchers could now develop a tool for rapidly analyzing the genome, to home in on regions of DNA that seem to be associated with disease: the genome-wide association study, or GWAS.

The first step in a successful GWAS is to collect as many DNA samples as possible from people diagnosed with an illness of interest (for example, schizophrenia) and likewise to collect samples from a large number of apparently healthy people without that illness—the more samples, the better. With computer assistance, the GWAS method compares the information of these two groups, looking for any markers that are far more common among the people who are ill. Upon making this comparison, the theory goes, the genetic marker for any disease should be unmasked, almost instantly, for all to see.

In the first decade of the new millennium, there was a GWAS, and often more than one, under way for practically every disease suspected of having a genetic source: heart disease, diabetes, rheumatoid arthritis, Crohn's disease, bipolar disorder, hypertension. In 2005, DeLisi was chairing a meeting of the International Society of Psychiatric Genetics in Boston when Edward Scolnick, a researcher from the Broad Institute of MIT and Harvard, announced that his institution planned to become the world's clearinghouse for genetic data on schizophrenia, with an aim toward identifying schizophrenia genes with a GWAS. By 2008, virtually every researcher in the field, including DeLisi, participated in a new group called the Psychiatric GWAS Consortium (now the Psychiatric Genomics Consortium), which collected some 50,000 DNA samples from people with a range of psychiatric conditions, including DeLisi's samples from the Galvin family. And in 2009, using information from that consortium, a study of 75,000 irregularities from more than 3,000 people with schizophrenia and bipolar disorder revealed "thousands of common alleles [possibly mutated genes] of very small effect."

This psychiatric GWAS was finding potentially relevant genetic locations all over the place, suggesting a new and deeper understanding of how mental illness operated in the brain. In the years to come, this new knowledge would help geneticists see how schizophrenia and other mental illnesses significantly correlated with copying errors—or copy number variations (CNVs)—in which whole chunks of DNA are either overproduced or go missing altogether. But for those hoping that the GWAS approach would find just a few genes to pin the blame on, this was hardly encouraging. And this was just the beginning. The schizophrenia GWASes that followed identified the first several genetic locations that seemed especially relevant to the illness. One GWAS, published in *Nature Genetics* in 2013, included about 21,000 genetic samples and found 22 such locations. Another GWAS, published in *Nature* in 2014, involved 36,989 patients and found 108 locations. Robert Freedman's CHRNA7 gene was in one of these suspect locations, which offered him some nice outside validation. But the more they found, the less meaningful the results seemed.

Each of these genetic irregularities, taken by itself, only accounted for a minuscule increased chance of an individual having schizophrenia. The researchers tried to make lemonade out of lemons by considering all of these insignificant factors together, combining them to come up with what they touted as a "polygenic risk score." But to many researchers, the polygenic risk score was merely a lumping together of trivialities into something only slightly less trivial. The genetic markers identified in the 2014 *Nature* GWAS, taken together, would only increase one's chances of having the disease by about 4 percent. "It's sort of a mindless score," said Elliot Gershon, DeLisi's old boss at NIMH, who had moved on to the University of Chicago a few years after DeLisi left. "You can't really tell anything from the polygenic risk factor."

The GWAS approach was not delivering the tidy ending that geneticists like Varmus had expected. In the face of blistering disappointment, the leaders of the Broad Institute, which had been leading the GWAS efforts for schizophrenia, decided to double down—resolving to build a bigger and better GWAS. "The guess among my colleagues is that we'll need 250,000 schizophrenia patients," said Steven Hyman, the head of the Broad Institute's Stanley Center for

Psychiatric Research, "which is daunting, but feasible for this disease." By the time they were done, Hyman predicted, "there will be thousands of variants in many hundreds of genes" all pointing toward schizophrenia.

Some suspected that the entire process might be leading the field astray—sending researchers once again to go look for their lost keys where the light was, not where the keys really might be. After all that work, the underlying nature of schizophrenia remained a matter of intense debate. "Is it a classical organically based biomedical disorder," the psychiatric geneticist Kenneth Kendler wondered in 2015—like, say, Alzheimer's disease was thought to be—"or is it the severe end of a spectrum of syndromes that aggregate together in families?"

Lynn DeLisi knew where she stood on the matter. She'd known for years. "My thought was, 'I don't believe that these hundred genes or markers are going to lead to anything,'" she said. "I want to see what's causing schizophrenia in these large families like the Galvin family."

WHEN PFIZER PULLED the plug on DeLisi's research into families with schizophrenia in 2000, she was forced to stop all of her work. Like the parties in a divorce, she and Pfizer divided her physical samples from her families straight down the middle. The term in research is "aliquoted": She and Pfizer each walked away with one half of each of her blood samples, enough material, in theory, for both parties to continue work. But in a cruel irony, no work would continue: DeLisi had the will to keep going but not the money, while Pfizer had the money but lacked the will.

Why would any large pharmaceutical company not want to try to develop a better drug for schizophrenia—one that might hit a genetic target and resolve issues that Thorazine and its offshoots never could touch? The reasoning at the time, according to professionals dealing with those companies, was pretty clear. Even with a genetic target, like Freedman's $\alpha7$ receptor, the pipeline to develop and test such a drug was extremely expensive, requiring human subjects willing to endure unpredictable side effects. Which would be all right if there was a likely financial benefit at the end of that pipeline. As it stood, Thorazine and its offshoots had been around for so long that virtually every company had its own version; these drugs were so stable and so effective at soothing psychotic episodes that it was hard to financially justify spending money to develop something new.

DeLisi called what happened with Pfizer "a disaster." With no other options, she kept her half of more than a thousand blood samples from three hundred families—including her half of each of the samples from members of the Galvin family—in a freezer at her new position at New York University. After a blackout hit New York City in 2003, DeLisi gave her samples to a colleague at another institution for safekeeping—first Cold Spring Harbor, then the University of California in San Diego.

DeLisi's family samples weren't so much gone as they were in exile. She had no idea how long it would take to bring them home again.

DON
MIMI

DONALD
JIM
JOHN
MICHAEL
RICHARD
JOE
MARK
MATT
PETER
MARGARET
LINDSAY

CHAPTER 34

After his time in Boulder with Lindsay, Peter returned to the same revolving door between Pueblo and Hidden Valley Road. His sister gave
up her legal guardianship so that Peter would be treated as a ward of
the state, allowing him to stay for long periods at state hospitals if
needed. His current diagnosis was bipolar disorder, rounded out by
the occasional delusional episode. For a decade, each hospitalization
lasted just long enough to get him out on his own again. Each trip
into the real world lasted only as long as he took his meds.

By 2004, he was forty-three years old, more ragged around the
edges, thinner, more addled. On February 26, after a two-month stay
at Pueblo, he was released on Risperdal, an antipsychotic, and Depakote, an epilepsy medication that also works as a mood stabilizer for
bipolar patients. He had not been taking either when he was readmitted three days later, on February 29, convinced that George W. Bush
was bombing the Broadmoor Hotel in downtown Colorado Springs.
This would be his twenty-fifth admission to Pueblo.

This time, the doctors gave him three different neuroleptic drugs,
Thorazine every two hours and two atypical neuroleptics, clozapine
and Zyprexa, twice daily. Once a day he was also given Neurontin,
an anti-seizure drug sometimes prescribed for alcoholism. Nothing

seemed to work. In April, two female patients on the ward said he grabbed and kissed them. In June, he was spotted purposely throwing up his medications in the bathroom. Over the summer, he lunged at a hospital staff member, pounded on the walls, and called other patients "pussy," "bastard," and "asshole." To one staffer, he said, "Don't you come near me with your medication, you bitch"; to another, he said, "I am going to kill you." Peter grabbed the phone receiver while fellow patients were on a call and hung up, turned off the TV while others were watching, flooded the bathroom. And he began preaching to everyone around him. "I am Moses. You will burn in hell. Take your clothes off. You are all lepers. You're dead. I will take a bat to your head. Shut up, or I will fuck you up." More than once, he received the ultimate disciplinary measure at the time: seclusion and restraints. By August, Peter's regimen had expanded to include eight different drugs: Geodon, Risperidone, Neurontin, Risperdal Consta (an injectable drug), Zyprexa, Prolixin, Trileptal, and Thorazine. That, too, didn't work.

And so, on September 14, for the very first time, after the doctors acquired the proper court order, Peter started a course of ECT—better known as shock therapy.

NO PROCEDURE FROM the dark ages of mental illness treatments has experienced as unlikely a cultural rehabilitation as electroconvulsive therapy, or ECT. The use of electricity to induce a seizure and calm the brain had been a cultural shorthand for medical torture for decades—since, perhaps, Ken Kesey made it the climactic stroke of barbarism inflicted on McMurphy in *One Flew Over the Cuckoo's Nest*. By the time Peter first had the treatment, however, a fine-tuned version of the same technique was being described as effective, safe, and even relatively painless. ECT's ability to nip mania in the bud with bipolar patients is so well documented that it was a matter of time, perhaps, before Peter became a prime candidate.

Everything about this new, improved ECT seemed designed to counter all that had been said about it decades earlier. Patients are sedated when receiving the shocks. They're given a muscle relaxant to reduce anxiety, and everything happens while they're asleep. The procedure still is known to have adverse effects on patients' memories, particularly after many treatments. And yet in some cases ECT seems to be able to adjust serotonin and dopamine levels more effectively than any medication. There are many stories now of accomplished, talented people—Vladimir Horowitz, Senator Thomas Eagleton, Thelonious Monk, Carrie Fisher, and Dick Cavett among them—using ECT to right themselves, usually with just a few treatments or even just one.

What happens if you need more than just a few treatments is another matter. Would Peter lose his memory, his sense of self, his personality? Risk aside, the decision was no longer Peter's to make. He was a ward of the state now, and his doctors had the ability to petition the court on his behalf. Mimi, assuming she was consulted at all, was not inclined to contest any decision the doctors made. This may be the only thing that actually helps Peter, she'd say. How much worse off might he be without it?

ON THE SECOND floor at Pueblo, Peter would change into scrubs, lie on a table, and receive general anesthesia. His mouth was covered, a machine helping with his breathing. He was given a caffeine tablet to lower his resistance to seizures—allowing the doctors to use less voltage—and a drug called Robinul that would prevent drooling. There was no arching of the body, no movement at all, except, perhaps, the jaw. When he awoke, Peter would be groggy. He'd get more caffeine, a pill or some coffee, to clear his head.

Peter never liked it. "I'm not having that shit," he said on November 8, 2004. "It messes with my bones. I'm calling the Air Force Academy and having them bomb this place."

He was placed in seclusion with restraints many times in December, once for forty continuous hours. He was still receiving ECT once a week, and he continued to throw up his defenses at every turn. He told one staffer, "You're a bitch. You will be fired if you mess with me and my attorney. I am suing you for fifty billion trillion million dollars. . . . You are whores of Babylon. . . . My arm broke last night, but I healed it."

When one hospital staffer told him not to drink any fluids the night before his ECT session, he said, "Fuck you, bitch. I can do what I want. You are going to die." That same month, he kicked a hospital worker in the side, breaking her rib.

It took more court orders for the mental hospital to get the authority to administer more treatments, but once they increased the frequency, the doctors noticed a difference. Peter averaged three ECTs a week for three weeks and then twice a week through January 2005. Finally in May, he was declared symptom-free. "He exhibited no evidence of dangerousness. He had earned privileges. He had gone on pass with family members and had done well," his discharge document reads. "The problem is that Mr. Galvin continues to lack insight into his illness. . . . Mr. Galvin has not bought into the idea that ECT is all that necessary for his future stability, or something that he envisions continuing on an indefinite basis." It was for this reason that the doctors designated his long-term prognosis as "guarded."

A year later, in June 2006, Peter was back at Pueblo, declaring, "I had a restoration of the spirit." He refused to eat because he believed his food was poisoned. He had been talking to Jesus. He was Saint Peter, and the devil was after him.

The staff at Pueblo scheduled more court hearings to maintain the frequency of his ECT sessions—once a week or even more, as needed. Peter didn't want this. But it wasn't up to him; it never had been. As he said in one intake meeting—asked, one more time, to recite his medical history—"Mental Health got ahold of me and ruined my life."

CHAPTER 35

Jim Galvin had been in and out of the emergency room at Penrose Hospital in Colorado Springs for weeks, complaining of headaches and tingling in his extremities. The staff sent him home again and again, writing off what he was saying as signs of his usual paranoia.

Toward the end, Jim believed he had a hole in his chest. "Don't you see I've been shot?" he said.

Jim died alone in his apartment in Colorado Springs on March 2, 2001, at the age of fifty-three. The doctors recorded the death as heart failure, related to his use of neuroleptic drugs. His family took this to mean that he died of a condition called neuroleptic malignant syndrome—a rare, life-threatening disorder most often caused by the drugs meant to help. Researchers predisposed against the reflexive use of medication to treat the mentally ill have attributed tens of thousands of deaths to this syndrome. Some of the symptoms, like agitation and delirium, can easily be mistaken for psychosis, which explains why the syndrome is often only identified after the patient dies. Still other symptoms, like cramps and tremors, are often the same as the side effects of the drugs. Jim's symptoms were so pronounced that at the time of his death, he had been prescribed procyclidine, a drug usually used to curb the effects of Parkinson's disease.

For Margaret and Lindsay, the lesson of Jim's death was clear. The cure was as bad as the disease. The sisters looked around and saw four more brothers—Donald, Joe, Matt, and Peter—and wondered who might be next.

Mimi, meanwhile, continued to explain Jim's mental illness in terms of his life circumstances—a bad marriage, and maybe even the trauma of being made into an abuser by a nefarious priest. She was not ready to cast any of her children aside completely, not even Jim, despite what her daughters had revealed about him. "It was the strain of the marriage, I think, as much as anything," Mimi would say, "and probably his own guilty conscience. But he was so well liked by *all* the children."

Jim's ex-wife, Kathy, and their son, Jimmy, did not attend the funeral. They were living in California now, rebuilding their lives, trying to forget the man who had tormented them both.

THERE WAS ONE question about their childhoods that Margaret and Lindsay had left unasked until now. With Jim gone, they saw an opening. Their parents had known that Jim was mentally unstable— even hospitalized—early on in his marriage. Why had they allowed both girls to spend nights in Jim's home, weekend after weekend, alone with him?

One day in 2003, tape recorder in hand, Margaret put the question to Mimi, point-blank. "Why did you let me go to his house that whole time?"

Mimi answered at once. "Because he had a recovery," she said. "He had a recovery. He went back to work. His wife took care of him, and everything seemed fine. And he had subsequent breakdowns—he was seeing an outside doctor—and he would recover and be fine for six months."

Margaret's voice broke as she responded, her voice faint as a child's.

"No one ever told me he was sick."

"Oh my God," Mimi said—less shocked than exasperated about having to rehash this again.

"I never knew," Margaret said.

"See, they didn't *know* in those days, Margaret," Mimi said, speaking quickly now. "You know, they snap right out of it, it seemed. And he did. He came back, he would go back to work and hold it together. But he always overdid. He'd not just do the one job, he'd try to do two. And he was working like eighteen hours a day, things like that, and so he'd have a collapse. And he was drinking. Right. He was drinking."

———⁓———

When cancer finally took Don Galvin—on January 7, 2003, at the age of seventy-eight—he had wasted down to less than a hundred pounds. He received a funeral with full military honors at the Academy Chapel, the architectural showpiece of the Academy that he had helped to launch into the world. One of the Academy's roster of performing falcons was in attendance, perched on the fist of an Air Force cadet for the entire length of the service.

Michael played classical guitar as people entered. The boys' old piano teacher played "Be Not Afraid" as the opening hymn. Mark, the former chess prodigy, read aloud from the description of a wise man in Chapter 39 of Ecclesiastes that fit well enough with the way Don wished to be seen in the world: *He researches into the wisdom of all the Ancients, he occupies his time with the prophecies / He preserves the discourses of famous men, he is at home with the niceties of parables. . . .*

Richard read from the book of John. John, the music teacher, in from Idaho, conducted the prayer of the faithful. Donald—nearly sixty now—read from the Beatitudes about the four virtues: prudence, justice, fortitude, and temperance. Michael and Lindsay each read poems. Margaret delivered a eulogy. "His memory failed him late in life," she said, "but that didn't mean his life was ordinary. It was extraordinary."

The service closed with "On Eagle's Wings." The Air Force's Thunderbirds were scheduled to conduct a flyover in Don's honor over the grounds of the Air Force Academy—the place where he had been happiest—but foul weather grounded the planes.

Don's body was sent to the University of Colorado, where Robert Freedman's team examined his brain. They were surprised to find that

Don's brain did not have any of the physical attributes that were associated with mental impairment or illness.

Mimi had little to say about that. She knew what she knew.

—⚭—

The last time Joe, the mild-mannered seventh son, living alone in his Section 8 apartment, talked to his mother, he told her that his feet were numb and that he couldn't walk. It was snowy, and Mimi couldn't drive in bad weather. She said she would see him in the morning. By then, it was too late.

Joseph Galvin died alone at home on December 7, 2009, at the age of fifty-three. The county coroner's report marked the cause of death as heart failure, caused by clozapine intoxication. The powerful atypical neuroleptic had proven helpful to Joe in some ways, but the drug's physical side effects seemed to slowly wear his body down. The echoes of Jim's death were unmistakable. Here was another likely instance of neuroleptic malignant syndrome.

When Jim had died, neither Lindsay nor Margaret had been tempted to mourn him. Joe was different. As children, both sisters had fantasized about not having brothers. The truth was, seeing the brothers they loved like Joe develop schizophrenia felt a little like watching them falling off the face of the earth. And so when Joe really did die, it was hard to grapple with what seemed like both a loss and the echo of that other loss, experienced years earlier, when mental illness took him away.

The family gathered to scatter Joe's ashes. Peter's face was ruddy, his clothes shabby and smelling of cigarettes, but he was still boyish, his bright blue eyes twinkling, his hair still jet black. Donald told the group that when he died, he wanted to be eaten by an elephant. Michael wanted his ashes scattered, but he wasn't sure where quite yet. Richard had a place in mind: Boreas Pass in the Rocky Mountains. Margaret said she'd like to be in Maroon Creek in Aspen. Lindsay picked the back bowls, a skier's paradise at Vail.

They all reminisced about their best times with Joe. Donald mentioned watching him play hockey. Mark remembered Joe racing his GTO against a Datsun 240Z and winning. Peter recalled living with him briefly in Chicago, when Joe was still throwing bags for United.

Margaret talked about him teaching her how to drive a stick shift, off of Arapahoe Road.

Mimi went further back in time than anyone else—reflecting, perhaps, not just on the boy Joe had been, but about the time of her life when all the boys were still young, and when happiness still meant the promise of something wonderful to come. When he was a baby, she said, Joe was so beautiful while he slept. Like an angel.

CHAPTER 36

2009

Cambridge, Massachusetts

In 2009, Stefan McDonough was entering his seventh year at Amgen, a neurobiologist lured out of academia by the prospect of developing real-life treatments and cures for one of the world's largest biotechnology companies. After a few years researching new pain-management drugs, McDonough's portfolio in the neuroscience department had broadened to diseases of the brain, including schizophrenia. Amgen was looking for a gene that could be targeted, something that needed rejiggering to help people with schizophrenia; if McDonough could find such a gene, Amgen would set to work on developing a drug to attack it.

From his office in Cambridge, McDonough threw himself into the work. He was so enthusiastic about the genomics revolution's potential that he arranged to audit an undergraduate genetics course at Harvard, sitting after work in an old wooden one-armed desk chair, week after week, dreaming of finding the gene that would prove to be schizophrenia's smoking gun. But very quickly, McDonough grew frustrated. Despite all the fanfare, it was clear every genetic location that had been associated with schizophrenia since the completion of the Human Genome Project—and there were more than a hundred of them now—had an effect so tiny that the idea of making a drug targeted at any one of them seemed ridiculous. That was when McDonough started looking around for another way—a shortcut to narrow the search. Wouldn't it be easier, he thought, to find a smaller haystack to rifle through? Instead of searching the genetic code of many thousands of unrelated people, why not study a limited group of people who had seemed to inherit the disease because of a genetic irregularity they all shared?

Why, he thought, wasn't anyone researching families?

McDonough was hardly unaware of the drawbacks. He knew that one family's genetic mutation—or, as the field now calls it, "disease-causing gene variant"—could be unique to that one family, and pointless to spend resources on. And yet he also knew that one family's abnormality might reveal something fundamental about the illness that everyone had been missing. He needed to find someone who felt the same way—an expert on schizophrenia and families who could teach him more. He found a professor at Harvard, an easy stop on his commute home from work in Cambridge, who was kind enough to talk about imaging the brains of schizophrenia patients. Families weren't her thing.

But she did know Lynn DeLisi.

"I'VE GOT MY name on more papers than I need," DeLisi said. "I just want to find these genes and help solve this disease."

She was working not far from McDonough, in Brockton, where she had just joined the staff at the VA Boston Healthcare System's psychiatric facility. That very year, 2009, she had moved from New York to Massachusetts, where she also was teaching classes at Harvard Medical School. Since her split with Pfizer in 2000, DeLisi had been estranged from her own research; no company seemed interested in picking up where the sale of Parke-Davis had left her, until now.

As she listened to McDonough talk about what he wanted to do, it was hard to say what she felt more intensely—surprise that a pharmaceutical company was interested in her work after all this time, or impatience to get started again. For McDonough, DeLisi checked off every box: a world-class researcher who had broken ground in this field; a devoted clinician who cherished one-on-one interaction with patients; a determined geneticist who yearned to find a cure. Best of all, she had been collecting pedigrees of families with schizophrenia since before McDonough graduated high school. And she was nice—something McDonough appreciated, given how territorial and guarded some academic researchers can be around pharma people.

She invited McDonough to join her on rounds at the VA hospital's inpatient psych ward. This would be the biotech researcher's first face-to-face contact with people suffering from the condition he wanted to cure. He watched as DeLisi, soft-spoken but direct and firm, visited with one patient who seemed perfectly calm, his delusions controlled, only to learn later that the man had committed an unspeakable crime. Other patients at the VA seemed completely soothed by the medication, but matter-of-factly noted that yes, they were hearing voices. "They're telling me to kill people," one said.

McDonough began to see how many of the patients were cognitively present, but without the emotions that typically make a person seem truly there. Only when he finally saw one patient having a violent breakdown, barricading his room and hissing with rage at the attendants around him, did he understand the plight of everyone there. "They have been warehoused where nobody can really deal with them," he said.

Here was the real reason, he thought, why big pharma could afford to be fickle about finding new drugs for schizophrenia—why decades come and go without anyone even finding new drug targets. These patients, he realized, can't advocate for themselves.

DeLisi made a deal with Amgen to work with McDonough on a new schizophrenia study. There was a maze of bureaucracy to contend with. First came a question of whether DeLisi was, in fact, the owner of her family samples, since the work of collecting most of them had taken place while she was with an institution, SUNY, that she since had left. Next, Amgen needed documentation showing that every donor in DeLisi's collection had consented that their biological data could be used for research. Hundreds of emails later, she and McDonough finally retrieved a selection of samples that DeLisi had stored at the Coriell Institute—DNA from some three hundred families, faithfully preserved in culture.

When McDonough got his first look at DeLisi's old research, he was astonished. The complete sequencing of a genome had been impossible in the 1990s, and yet the level of analysis she had done on these samples was ahead of its time. Now, these samples had woken, Rip Van Winkle–style, in the age of computer-assisted genetic analysis. The analysis would be easier than ever now—and more precise, more nuanced, more detailed.

FOR THIS NEW study, they wanted only the most blatant, egregious multiplex family cases they could find. At least three people in each family had to have schizophrenia, and at least three others in the family must not. They settled on nine families—four whom DeLisi had contacted while she was at the VA, and five from DeLisi's old collection. The Galvins were from the latter group, the largest group by far of siblings of any family in the sample.

McDonough and DeLisi's goal was to see if any of these families carried a rare genetic mutation or irregularity that was shared by the sick family members. This was what made the size of the families in their analysis so important: Any gene variant found in a person with schizophrenia, DeLisi and McDonough knew, might also be present in an affected parent or sibling simply by chance—and not because it is a cause of their shared disease. Parents share half of their genes with each child, after all, and a variant present in one sibling has a 50 percent chance of being present in a brother or sister. But as the number of family members with schizophrenia grows, it becomes increasingly meaningful if each and every one of them has a specific gene variant. The likelihood that the mutation is harmless or unrelated to their schizophrenia dwindles and, as it tracks faithfully with the disease in more and more family members, the mutation looks increasingly likely to be the cause. The assumption they were making was that any rare mutation they found would offer a fresh way of understanding the illness. "Even though that particular mutation may be unique to that family," DeLisi said, "it's possible that the abnormality in that gene is part of an overall biochemical pathway that may be abnormal in schizophrenia."

Sure enough, with the Galvins, DeLisi and McDonough found something tantalizing: a mutation shared by every Galvin brother for whom DeLisi had collected samples back in the eighties, in a gene called SHANK2. The mutation they found was connected to an important process in the brain—a process that seems vitally related to schizophrenia. SHANK2 is a communications assistant for brain cells. The SHANK2 gene encodes the proteins that help brain synapses to transmit signals and the neurons to react quickly. The Galvins' mutation significantly alters the protein that SHANK2 produces. "The mutation was found right at one of the known functional guts of SHANK2," McDonough said, "right at one of the spots that's known to be critical for SHANK2 function." In this way, the SHANK2 mutation was pointing the way to something new, potentially, about the illness—a glitchy molecular process that might be shared by more people than just this one family. Schizophrenia might take shape in that process. "It's certainly not a proof, to a scientific standard, that this mutation caused the disease," McDonough said. "What it really tells you about is a mechanism of schizophrenia."

Similarly rare variants have shaken up the research into other diseases. Parkinson's researchers, for example, found a genetic mutation affecting the α-synuclein protein in one family in Italy that pointed the way toward developing new drugs. The greatest example may be the development of statins, the drugs that decrease cholesterol levels for thousands of people at risk for developing heart disease. Scientists knew for years that high cholesterol contributed to heart disease, but no headway was made in lowering cholesterol until two researchers at the University of Texas Southwestern Medical Center in Dallas found some families with very early onset cardiovascular disease who had unusual mutations that impaired the body's ability to remove LDL cholesterol from the blood. This mutation was not present in most people with heart disease. But that didn't matter—the study of those mutations exposed how cholesterol levels can be lowered, not just in those families but in almost everyone. The drug that was developed to correct that particular LDL issue turned out to revolutionize the treatment of heart disease.

This may be the real miracle offered by the Human Genome Project: not the chance to find a smoking-gun gene that may or may not exist, but the ability to see how schizophrenia takes shape in the brain. SHANK2 is just one example of this; the way Robert Freedman's CHRNA7 gene shed a light on information-processing issues is another. And at the same time that DeLisi and McDonough were doing their work, a team from the Broad Institute in Cambridge, the Harvard-MIT collaboration that had taken charge of the GWAS efforts for schizophrenia, published its own highly publicized study identifying a mutation in a gene called C4A—more common than the mutation in SHANK2, but still far too rare to target with a drug— that seemed to play a role in the overpruning of brain synapses. Their research suggested that people with schizophrenia might end up cutting some synapses as adolescents that they would need later in life— another angle on the process of schizophrenia. While it is not clear if the Galvins have that C4A mutation, they played a small role in that study, too, being among the earliest families to donate their DNA to the pool of data analyzed by the Broad Institute's team.

DeLisi and McDonough's study was published in *Molecular Psychiatry* at the end of 2016. It was not possible to say for sure that this particular SHANK2 variant in this specific gene caused the Galvins' schizophrenia. But that conclusion was consistent with what DeLisi and McDonough saw. Thirty years after she first met the family in their living room in Woodmen Valley, DeLisi had arrived at what looked like an answer to the question that beset the Galvins: Why?

THAT ANSWER CAME with some surprises. The first involved the connection of the genome's three different SHANK genes— SHANK1, SHANK2, and SHANK3—not just to schizophrenia but to other mental illnesses. Before this study, others had conducted separate studies of each of the SHANK genes' relationship with autism and other brain disorders. Now, taken together, all the research indicated that at least some varieties of mental illness exist on a spectrum: Some people with certain SHANK mutations may have autism, while others are bipolar and still others have schizophrenia.

The concept of a spectrum of illness seemed highly relevant to the Galvin family. Peter, for example, wandered between diagnoses, from schizophrenia to bipolar disorder. Donald also was diagnosed with mania and prescribed lithium early on, before the doctors moved on to the usual assortment of neuroleptics. Joe's collection of symptoms was different from Jim's, and Jim's was different from Matthew's— and surely there was no one else like Brian. Yet seven of the brothers— the seven who provided DeLisi with samples, including at least a few nondiagnosed brothers—all had this same mutation, in a gene that also figured prominently in other mental illnesses.

"Lynn was right," McDonough said. Studying families with multiple occurrences of mental illness was, in the end, the study of a shared genetic issue—one that, depending on each person, manifests itself in a different way. "These are multiplex families, and it sure looks like the same genetic determinants can give rise to subtly different diseases."

It's possible that discoveries like the Galvin family's mutation could point the way toward a completely new conception of mental illness. That could come sooner rather than later; in some corridors, it's already happening. In 2010, the psychiatrist Thomas Insel, then director of NIMH, called for the research community to redefine schizophrenia as "a collection of neurodevelopmental disorders," not one single disease. The end of schizophrenia as a monolithic diagnosis could mean the beginning of the end of the stigma surrounding the condition. What if schizophrenia wasn't a disease at all, but a symptom?

"The metaphor I use is that years ago, clinicians used to look at 'fever' as one disease," said John McGrath, an epidemiologist with Australia's Queensland Centre for Mental Health Research and one of the world's authorities on quantifying populations of mentally ill people. "Then they split it into different types of fevers. And then they realized it's just a nonspecific reaction to various illnesses. Psychosis is just what the brain does when it's not working very well."

THE SECOND SURPRISE was about Mimi. For decades, Mimi had insisted that the family illness came from Don's side. As far as she was concerned, his history of depression proved it, and no one who researched the family ever had reason to disagree with her. "We were looking for something transmitted from the father," McDonough said.

The SHANK2 mutation, however, came from the mother's side of the family—suggesting that it could have been Mimi all along who was the carrier of the mutation responsible for the family illness. Another study of SHANK2 and schizophrenia, published at about the same time as McDonough and DeLisi's study, noted several more instances of an unaffected mother passing along a mutation to a son who developed the disease. It should also be possible for a father to be an unaffected carrier—SHANK2 is not a sex-specific gene; it is not located on the X or Y chromosome, which determine sex, but on chromosome 11.

Why did six out of ten boys develop serious mental illness, but neither of the two girls? It might simply be chance—a roll of the dice working out for both sisters and four of the ten brothers. It could be that the Galvins' SHANK2 issue points to, as DeLisi and her coauthors suggested in their study, an "as yet uncharacterized sex-dependent influence" on how the illness develops—though that would not account for the Galvin sons who *didn't* get the disease.

Or it could be that the mutation on the mother's side is mingling with something else on the father's side—that the SHANK2 mutation does nothing by itself, but needs another mutation elsewhere to completely set the table for the disease. Such is the way, sometimes, with genetic mutations. The geneticist Kevin Mitchell has noted how specific mutations can manifest differently in different people: The same mutation can trigger epilepsy in some people while in others it triggers autism, schizophrenia, or nothing at all. And sometimes, a second rare mutation elsewhere in their genome suggests a combined effect.

It is possible, maybe even likely, that the genetic flaw that caused schizophrenia in the Galvin boys might not be Mimi's fault or Don's fault, but both of their faults together—an entirely original cocktail, powerful enough to change all of their lives.

CHAPTER 37

2016
University of Colorado Medical Center,
Denver, Colorado

While DeLisi and her new partners at Amgen followed the trail of the SHANK2 gene in Cambridge, Robert Freedman was continuing his work in Denver. Like DeLisi, Freedman had experienced a long period of early promise, followed by a painful reversal of fortune—the excitement of isolating the first gene to play a confirmed role in schizophrenia, and the anguish of watching the drug trials to activate the brain receptor for that gene go nowhere. Freedman had hit a wall, and now he was searching for another way in—a new strategy to help repair or strengthen the one gene he knew made a difference.

One thought he kept coming back to was that when it came to his prized gene, CHRNA7, researchers might have arrived too late to help adult patients like the Galvin brothers. Like many genes, this one is fully developed in the womb, before birth. Freedman tended to think of the development of a baby's brain as a series of computer upgrades: A fetus starts with a very simple operating system, and as it grows, that operating system installs the next, more sophisticated system. The CHRNA7 gene appears early in utero; its purpose, as far as Freedman could tell, is to help install the final OS, the one that we use as adults. This would mean that by the time a baby is born, the die is cast. If he was right that schizophrenia hinges on the condition of CHRNA7, the only option might be to try to fix it before birth.

Freedman's task seemed clear: If he could repair flaws in CHRNA7 in the womb, he would have a chance at nipping schizophrenia in the bud, before the disorder ever materialized. If he could manage that, then Freedman might be able to successfully keep an entire generation of people genetically predisposed to getting schizophrenia—and all the generations after that—from ever becoming symptomatic. He could hardly imagine a less realistic goal. The Food and Drug Administration would have to agree to an experimental drug for women who were pregnant, and that seemed, to say the least, unlikely. Medicating fetuses—drugging unborn babies—just wasn't going to happen.

What Freedman needed was a method that would not involve surgery or synthetic drugs. What he found, somewhat miraculously, was that acetylcholine—the substance that runs the particular information-processing operation of the brain he wanted to target—is not what CHRNA7 needed the most when it was first getting started. What this gene really needed during the fetal stage was a nontoxic, utterly benign nutrient available at every GNC and Vitamin Shoppe in America.

Choline is in a lot of foods people eat every day, including vegetables, meat, eggs, and poultry. Pregnant women dispense choline to their unborn children as part of their daily diet, through their amniotic fluid. Freedman's idea was simple: What if a mother-to-be of a child who is predisposed to developing schizophrenia took megadoses of choline, while her child was still in the womb? This could be a nutritional supplement, like the folic acid in the prenatal vitamins that pregnant women are encouraged to take to prevent spina bifida and cleft palate. Maybe then, an at-risk child's brain would develop healthily, in a way that it otherwise wouldn't.

The FDA agreed to an experiment. Freedman's team in Denver conducted a double-blind study in which some expectant mothers received high doses of choline. The women in the control group were observed to ensure that they ate enough meat and eggs, to make sure no one in the study went without at least an adequate amount of choline. When the babies were born, those who got the choline supplements in utero passed Freedman's double-click test measuring auditory gating: 76 percent of them had normal gating, compared to 43 percent in the control group.* Even babies with the CHRNA7 irregularity had, in more instances, normal auditory gating. The good news continued as the babies got older. At forty months, Freedman's team observed that the choline group had fewer attention problems and less social withdrawal compared with the control group. Choline seemed to work well on virtually everyone.

Freedman's study about choline was published in 2016, the same year as the Broad Institute's C4A study and DeLisi's SHANK2 study. In 2017, the American Medical Association approved a resolution that prenatal vitamins should include higher levels of choline to help prevent the onset of schizophrenia and other brain developmental disorders. It had taken thirty years, and he'd hit at least one dead end along the way. Only time can say for sure what difference choline might make over a matter of decades. But thanks in part to his work with the Galvins, Freedman had arrived at a game-changing strategy for the prevention of schizophrenia.

IN THE FALL of 2015, Freedman traveled to New York for an annual symposium sponsored by the Brain and Behavior Research Foundation, the group once known as NARSAD, which raised millions of dollars for research into new mental illness treatments. The results of Freedman's choline work had already made the rounds, and Freedman was there to receive one of the highest honors in his field: the Lieber Prize for Outstanding Achievement in Schizophrenia Research.

In New York to celebrate with him was Nancy Gary, who, along with her husband, Sam, had funded a chair in Freedman's psychiatry department at the University of Colorado. The choline study had captured Nancy's imagination. Years earlier, she and Sam had funded the construction of a pavilion at the university's hospital for children

---

* The double-click test, remember, is not a foolproof test for schizophrenia, but a measurement of sensory gating, which is just one of an unknown number of aspects of schizophrenia. Which is why 57 percent of a control group can fail the test, but still not have schizophrenia.

with psychiatric disorders. Now, with Nancy in her eighties and Sam in his nineties, they had pledged to support Freedman's next project, to trace children who'd received choline supplements in utero over several decades of their lives. They understood what might happen if choline really made that much of a difference: Some varieties of schizophrenia could go the way of the cleft palate. "The man is brilliant," Nancy said. "I would support him in whatever he does, because he's that good."

Nancy brought a guest with her on her plane to New York—someone Freedman hadn't seen in several years, not since she'd been living with her brother Peter in Boulder. Until Nancy reintroduced Lindsay to Freedman, the doctor had no idea of their connection. This felt like a *This Is Your Life* moment for him: One of his most generous benefactors had also been helping the largest family ever to have contributed to his research.

After many years in Boulder, Lindsay and Rick had moved to Vail, where Lindsay still ran her corporate event business and Rick worked as a ski instructor. Together they were raising their children—a girl, Kate, and a boy, Jack, both now teenagers. Lindsay and Nancy had fallen out of touch for several years, until one day they ran into each other on the slopes of Vail, where Jack was in the same ski group as one of Nancy's grandsons. Nancy was thrilled to reconnect with both Lindsay and Margaret, inviting the sisters to family gatherings in Vail. Mimi did not take part in the reunion—her days of socializing with Nancy Gary were long behind her—but both sisters were excited to be back in regular touch with the family that had made such a difference in their lives.

Lindsay brought her daughter with her to the symposium in New York. Nancy put them up at the Pierre Hotel, and they all sat together to watch Freedman deliver his acceptance speech. "Now, one of the unfortunate things about doing human research on the life cycle is it's the same as your life cycle," Freedman said, getting a chuckle out of the audience. "I'll be one hundred and thirty-five years old by the time we finish this study. A young investigator will have to call me in the nursing home and let me know if this works out just the way I planned."

Afterward, when the doctor had more time to talk, Nancy

beamed as she told Freedman that she'd been the one to send Lindsay to Hotchkiss—air-lifting both her and her sister out of the Galvin family home when things were at their worst.

Lindsay smiled silently, opting not to get into how technically it wasn't Nancy who paid her Hotchkiss tuition, and that there had been three years when her sister had been pulled out of that house by the Garys and she had not. And she smiled some more as Nancy began praising Lindsay and all she'd done. Here was the one she rescued, Nancy said, the one who survived—the Girl Who Lived.

MIMI

DONALD

JOHN

MICHAEL

RICHARD

MARK

MATTHEW

PETER

MARGARET

LINDSAY

CHAPTER 38

Hi Galvin Gang,

New news on the research front! Harvard study would like to take blood samples of Mimi and Papa's grand kids that are currently over 18 years of age. This is continued effort in research on schizophrenia.

Mimi and Papa have been active in research since the late 70's. Dr DeLisi will be sending a phlebotomist this fall. Thought we could make a celebration of it.

I know we all would like to find the cure for this tragic disease. Stand by for when and where to gather.

Love to All!

Mary

Email from Lindsay to family members,
September 9, 2016

On a Sunday in November, more than a dozen members of the Galvin family were invited to Margaret's house in Boulder. The purpose of the gathering was to collect as many DNA samples as possible from Galvin family members who might not be mentally ill, to use as controls against the samples the researchers already had. An assistant of

DeLisi's had flown in for the occasion, along with a phlebotomist who would bring the family's samples back to Boston.

"It's like a blood-drawing party," Margaret said. "It should've been on Halloween."

DeLisi had contacted the Galvins that summer, in advance of the publication of the SHANK2 results. Until then, no one in the family had the slightest idea that their blood samples formed the cornerstone of NIMH's research into the genetics of schizophrenia—a data set that has played a part in nearly every genetic study of the disease ever since. DeLisi had not been in touch with the family for decades— perhaps not since the late 1980s when a colleague of hers contacted them to follow up on DeLisi's first visit. Whichever family member took the call refused to set up an appointment and asked them not to call again. It happens sometimes: Families change their minds, or the researcher calls at an inopportune moment.

The family was not named in the study itself, of course. But DeLisi eagerly shared the news of the mutation with Lindsay, who spread the word to her sister and mother. Mimi, who had just turned ninety, was humbled a little by the news. She had spent so many years blaming the illness on Don's side of the family that there was little for her to say now, except to laugh shyly. But for both Lindsay and Margaret, there was no small amount of schadenfreude, seeing their mother so definitively disproven. And it was exciting for them, too, to see the research that had begun so long ago continuing now with the possibility of something to show for it. For the first time in years, they felt something like hope.

No one at the blood drawing would be told if they carried the SHANK2 mutation or not; the results of the DNA tests would be used anonymously, for research purposes only. "We're numbers, not names," Margaret said. Still, she and Lindsay were surprised to see who came and who didn't—those who wished to acknowledge the genetic issue they might carry, and those who would just as soon be- have as if the problem never existed. Their brother Michael came, but he was not pleased; it felt to him like pouring salt on an old wound. Mimi's sister, Betty, now in her late eighties, had married and had children who seemed free of the mental illness that haunted Mimi's family. She was still living on the East Coast, too far away to make the trip, and neither did her children or their families. The next generation had especially spotty turnout. Michael's kids came; Mark's children did not; and the son that Richard had fathered at the age of seventeen seemed put out even by the suggestion that he should be there.

The medical people told Lindsay that no-shows weren't so unusual. They're terrified of the disease, and simply don't want to think about it.

A WEEK LATER, Mimi inched from her bedroom, down a short flight of stairs, and over to a seat at her kitchen table on Hidden Valley Road. She moved confidently, but with the help of a walker with a portable oxygen tank hanging over the side.

"I am very arthritic and have replaced joints," she had said on the phone a few months earlier. "I'm like the Bionic Woman." She waited for a laugh, then said, "Not funny, dear. Wait till you get there. Two hip surgeries, and I'm ninety, and they'd like to do it again, but I'm too old. I'm simply worn out."

A clot in her eye made it hard for Mimi to read. "There's nothing like the feel of a good book in your hand, either," she said that day in the kitchen, "except that my hands are so bad now I can't hold a book." She had hearing aids in both ears that she fiddled with, struggling to understand people in groups. But she could still listen to a Salzburg recording of *Don Giovanni*. "When I'm here alone I can turn up the opera as loud as I want to, or a ballet, or whatever." Her mind was as sharp as ever; she remained stubbornly herself—intelligent and very well read, strong enough to have endured any number of horrible tragedies, and yet utterly averse to self-reflection.

Both sisters knew all too well how smoothly Mimi could change the subject when she wanted to, redirecting uncomfortable conversations whenever possible to her experiences with the Federation—"I could almost write a book about the people we met through that, the wonderful nights . . ."—or her teenage years exploring New York, or Don's military career. She took credit for the idea of making the falcon the Air Force mascot. "A lot of people have claimed they suggested the falcon first," she said, "but that isn't true."

Gently, the sisters tried to move her on to more meaningful mate-

rial, even if it meant making her uncomfortable. Though she did not
raise the subject, she answered a few questions about Nancy Gary, and
the years when she and Don and Nancy and Sam had socialized. "We
were quite close," she said. But she never had a one-on-one friendship
with Nancy. "Nancy has never been a buddy-buddy person, as far as I
can understand," Mimi said coolly.

"Why would I have gone to live with them, then?" Margaret
asked.

Mimi turned to her daughter. "Oh, because we had four kids in
the hospital at one—"

"I know—I know that side of the story," Margaret said. "But why
would they have taken me in, had you not been good friends?"

Mimi waved off the question. "I really don't know. Well, she saw
Brian had died, and she called."

THE MORE ON the spot Mimi felt, the more she leaned in to her
old perfectionism. "I don't paint anymore," Mimi said, glancing at
Margaret, "mostly because I can't compete with my daughter." Margaret had taken up painting, finally, with her own daughters older
now, and she was good. She chose natural subjects like her mother's
old paintings, but bolder, more inventive—and was even getting a
few sales, right away.

Then Mimi turned and looked at Lindsay. "*She's* always doing
something. She runs *beautiful,* big parties. But she lost that contract!"
She chuckled lightly. "She was giving the oil company party for a million dollars at a clip."

Lindsay kept a smile pasted to her face. As casually as possible,
she listed a few of her recent events, for an investment company and
a health care company.

"In twenty years, she's built up quite a clientele," Mimi said. "She
said, 'Mother, I won't do this very long, it's not very intellectual.' But
the pay was pretty good. She should have gone on to grad school!"

She turned to Lindsay. "Are you going to retire next year?"

"Hopefully," Lindsay said.

"Hopefully," Mimi echoed. "And then she's going to open a bookstore so she can read!" She gazed at her daughters.

"We both—all three of us—like to read," she said, beaming with
pride. "We're *all* readers."

———

NONE OF THE boys lived with Mimi anymore. Donald moved to assisted living at Point of the Pines three years earlier, after Mimi, sidelined for several months by a stroke, was too frail to care for him by herself at home. This saddened her. She liked having the company. But Mimi still saw them all, and she continued to snap at the sick boys, particularly Peter and Matt, whose hygiene appalled her. *Zip up your pants! Where's your belt? Go take a shower.*

Margaret and Lindsay understood that, to a point. But would things with the boys be any better if they were wearing ties and sport coats? By now, weren't digs like this beside the point? "She's not able to really share how she feels about any particular thing," Margaret said, out of earshot of her mother. "But she can be really critical about how the rice is being cooked."

At the kitchen table, the sisters both laughed.

"Mom," Lindsay teased, "if you had just said 'Yes' more often, there would be no schizophrenia."

Mimi's reply came quickly. "My problem," she said, "was I said 'Yes' too many times."

SLOWLY, WITH HER daughters next to her, Mimi was persuaded to talk about what had really happened—and how she really felt.

She remembered Jim, at the age of sixteen, threatening her with a large pot, and Donald trying to throttle her once when his medicine was misplaced. "It terrified me," she said. "If it hadn't been for three or four of the other boys, I think I'd be dead, because he really had me in a stranglehold."

She had no compunction about saying that Jim and Joe both died of the medicine that was supposed to help them. "Both of those boys would go to the hospital complaining of chest pain and get no attention," she said, "because they were mentally ill, and they both were dying of heart disease."

She recalled how shattered she'd been to learn about Father Freudenstein. She and Don suspected nothing, Mimi said, because who would? "Well, we weren't very savvy parents, not at all. We were as innocent as the day is long, and we let them go."

She talked about her husband's fragile mental state, which she felt was connected to his time in the war. "He saw a lot of action. But he never discussed it—I think he just kept it all inside." His hospitalization during his posting in Canada came ten years after the war. "The Air Force panicked because being an intelligence officer, they wanted him out of there quick. So he was brought to Walter Reed. No disease found. There was no test for PTSD."

When she talked about being blamed for her sons' mental illness, she got her back up again. "We were all involved in a discussion with the doctors," she said, "and they crucified us. We were the worst parents in the world. It made us feel terrible. It traumatized us. Don and I, we were both paralyzed mentally. It just freezes you, because you don't know what to do. You have nobody to talk to. We were an exemplary family. Everybody used us as a model. And when it first happened we were mortally ashamed."

She could talk about that shame now, unburdened at last. "Oh, that was the whole thing, it was so embarrassing. The blaming part really traumatized me to the point where I felt I couldn't tell a friend or anything. It was just all inside, and it was hard, that part. That's where I think the crutch of the Church kind of helped me. I was accepting it as my lot in life.

"And so I was crushed," Mimi said. "Because I thought I was such a good mother. I baked a cake and a pie every night. Or at least had Jell-O with whipped cream."

———

Between bouts of sympathy for her mother, Margaret remained highly critical of her—how everything in her life was centered around Donald and the other sick boys, to the exclusion of everything else, including the chance to have the relationship she wanted with her. "I never got to have my mom," Margaret said, "because of Donald." She viewed her grimly now, as a woman who sowed the wind and reaped the whirlwind. "She got her way," Margaret said, "and there was a large cost in that in terms of her relationships with her daughters and her other children who were not mentally ill. And so she really lost in the end. She pushed away the people who could have possibly had relationships with her."

In Margaret's view, that included her father. "I don't know, I'm not excusing his affairs, but I don't think there was a lot of thought going into any of this on either of their parts."

Margaret could imagine the scenario now in a way she couldn't when she was younger. Things got out of control because, with such a large family, the miracle would have been that anything ever was under control to begin with.

"I think there was a lack of consciousness in having twelve children in the first place," Margaret said, "and then thinking that they could raise them to become all-American citizens."

Margaret and Wylie had their own family now—two teenage daughters, Ellie and Sally—but the frightening aspects of her childhood were still alive inside her. Margaret never forgot how unsafe Donald and her other brothers made her feel as a girl, and so she would not see Donald on her own. She did not want him around her children, either. But now that she was married and well-off with a family, Margaret also felt guilty about having what many of her brothers did not. The act of buying high-priced leggings for her daughters at Lululemon could send her into a tailspin of self-judgment. Her sick brothers never had a chance at this kind of life.

They were going crazy, and I was swimming at the country club, Margaret wrote in her diary. *They are still crazy, and I am still swimming at the country club.*

So she tried half measures. Margaret helped from a distance, sending cash and gift cards, and propping her sister up on the phone, listening and sympathizing. But she still felt too vulnerable to be part of their lives. "It's like pouring a glass of water with no bottom in it. You can't ever fill it up. It's just futile to try and help them. It's not like they don't want to be better, but they just never get better. I really honestly stayed much farther away than Lindsay did." She stopped visiting the hospitals, and stopped bringing her kids along on the few times she did.

"I'm very lonely on the path through recovery from my family," Margaret said.

LINDSAY APPRECIATED HAVING her sister to talk with—"just knowing that someone else knows what you're talking about and knows the depth of the pain." But Margaret's distance from the family

now felt like another abandonment. She decided to do the opposite—
to continue caring for the brothers and seeing her mother and doing it
all. Lindsay ran point on every bureaucratic challenge her mother and
brothers faced: wrangling Social Security benefits, shopping for the
perfect housing situation, overseeing their medical care, advocating
for different medications when the current ones seemed to be failing.
She took over the boys' powers of attorney, and Mimi's, too. When
she assumed the caregiver role, she felt as if she were channeling everything about her mother that she admired—the tireless devotion
that both doctors, DeLisi and Freedman, had noticed when they'd
all first met.

"My parents were so devastated," Lindsay said. "My dad really crumbled. And my mother really transformed and became this
advocate."

Lindsay knew what she was doing was putting her on a collision
course with her sister. Where Margaret was staying away, Lindsay was
wondering why no one was helping do what so obviously needed to
be done.

"I'll work myself to the bone and not ask for help," Lindsay said,
"and then I'll be resentful."

———— ✺ ————

Lindsay was the first responder as usual when Mimi had another
stroke, in early 2017. Once Mimi was in the ICU, Michael and Mark
relieved her. Even Matt came by.

In March, Mimi was home and under hospice care, resting in bed
without wires or monitors attached. Unless you could afford full-time
help, hospice care didn't mean actual care, just supplies like morphine
and directions for how to care for your loved one yourself. In Mimi's
case, this meant dealing with incontinence and catheters.

Margaret joined them once Mimi was back home. She spent time
holding Mimi's hand, giving her little massages. Michael played Bra-
zilian tunes on his guitar. Lindsay tidied the house. The three of them
talked about old movies, enjoying being with one another. They spent
ten days this way, until, out of nowhere, Mimi started to eat again.

"I thought I was dying," she said. "That's why I wasn't eating."

Then she asked for a soft-boiled egg.

MARGARET HAD A trip scheduled to the West Coast, where her older daughter, Ellie, was touring colleges. She and Lindsay talked it over and agreed that Margaret ought to go, for Ellie's sake.

But on the day Margaret left, a confluence of events hit Lindsay in a way she was not expecting.

First her brother Matt—once her soccer coach, now on clozapine and living in government-funded housing—pulled up to the house in his jalopy.

Then came Peter—her old Boulder roommate, now an inpatient at Pueblo, receiving regular ECT treatment—driven in by Michael.

Then came Donald—her oldest family foil, the one she once dreamed of burning at the stake—driven in from his assisted living center by Mimi's housekeeper, Debbie.

All three sick boys were back at the house. Soon it would be just them, their mother, and Lindsay.

And Margaret was heading out the door.

Lindsay knew that this wasn't forever—that the brothers were just visiting. But none of that mattered. In a flash, Lindsay was ten years old—deserted, abandoned, forgotten, trapped. She tried as hard as she could, but the sensation shot through her like muscle memory: *It's happening all over again.*

IN THE WEEKS that followed, Margaret would come by for an hour or two, but not much more. Instead, she went ahead with a trip she'd planned with some friends to Cabo San Lucas in April, and from there headed off to Crested Butte, on vacation with Wylie and her kids.

Lindsay, furious with her sister, found herself fuming about any family member who didn't come to see Mimi. Mark lived in Denver, for God's sake—what was keeping him from driving to the Springs for the day? So did Richard, who had always been so attentive to Mimi—where was he now? Even John, whom she adored, had elected not to come back to see Mimi. He said he'd prefer to remember her the way he liked to remember her—that he didn't want to see her like this.

"They think it's weird that I'm hands-on," Lindsay said. "And I think it's weird you *wouldn't* be."

The exception was her brother Michael. In 2003, the hippie alum- nus of the Farm married his second wife, Becky, who went on to serve on the City Council in nearby Manitou Springs. Still wearing his hair in a ponytail, Michael assisted Becky with her horticulture business and still played small gigs at local restaurants—a completely healthy, functional life, with no psychotic breaks, no delusions, no schizo- phrenia. Michael's take on his sick brothers endeared him to Lindsay. "He thinks that traditional psychiatry has damaged them, which it has. I mean, there's no question," she said. Just looking at them— overweight, with tremors, stuck in their habits, unable to think of anyone other than themselves—you could tell they were no closer to cured than they were when they each had their first psychotic breaks.

Then again, Lindsay had tried everything else. "I don't know what the alternative is," she said. "I'm like, 'Well, Mike, if you're will- ing to take them into your home, off their medication, by all means, go for it.'"

Michael had experience in the hospice field. Over the years, he'd taken care of a man in Boulder, and his father-in-law, and his own father, Don, toward the end. Lindsay asked Michael to come and care for Mimi, sharing the duties with Debbie the housekeeper and a fam- ily friend, Jeff Cheney. All three, including Michael, were paid out of Mimi's account—a mixture of Don's military pension and some savings that Lindsay controlled.

Michael could use the money. But the chance to care for the woman who had loomed so large in his life made the job irresistible. He soon learned that however frail she might have been, Mimi was still in charge. He would offer her Kentucky Fried Chicken for din- ner, knowing how much she loved it, and she would refuse, saying she'd had it the night before. He'd make spaghetti instead, and she'd say there was too much of it.

"It got a little confounding," Michael said. "I almost dumped it on her head."

MIMI

DONALD
JOHN
MICHAEL
RICHARD
MARK
MATTHEW
PETER
MARGARET
LINDSAY

CHAPTER 39

"I have to—very slowly," Mimi said haltingly, her words slurring but her smile intact. "I've had I've had a crane—*brain*—problem. So my is very crazy. But you have to speak well and louder."

Half of the words that came out of Mimi's mouth were not what she intended them to be. She went back and forth for a full minute just on the word *Austria,* when what she really meant was *India.* "Most words came out as *water,*" Jeff Cheney, the family friend helping out as one of her caregivers, said. But she would not stop trying to explain herself, and was always chuckling a little.

"Margaret's here. She's how—*you* know. And my mouth is there— might be having to go, too—we'll see—but as I'm—you know—my original was eight dollar for being old, for getting too old."

Mimi tittered softly, exasperated. "Pretty bad. But I can try. Sometimes say *boy, school,* but today, *boy* or *book*!"

She laughed again. "So I'm trying a little. It's pretty bad. Not very good. And I thought I'd be, by now, I'd be *over.*" She laughed louder, and then out came something perfectly clear. "Well, as Mary said, 'Mother, you're just taking longer now!'"

The people around her had learned to decode much of what she was trying to say through her aphasia. They'd set up a hospital bed in the basement level of the house, easier for her caregivers to access. Each day brought something new: a bladder infection, an upset stomach, nausea, bouts of pain tempered by morphine. But Mimi could still watch TV—movies, cable news, and her favorite, Rachel Maddow. More helpless than she was accustomed to being, she would be alarmed when she was alone and go on tears about things she felt needed to get done around the house—most of them invented, like an overflowing septic system. For the first time in her life, Mimi had a few delusions of her own.

The longer Lindsay stayed, the more she understood her mother, or thought she did. When she wanted to get a complicated point across to Mimi, she would sometimes write her a note. When Mimi kept on refusing food and ordering something different, Lindsay wrote her, saying she believed these were her mother's final few attempts to try to control what was going on in her life. Mimi agreed with Lindsay, but she kept on doing it, anyway.

What Mimi could no longer do, thanks to the aphasia, was control the conversation. "This is my son," Mimi said, introducing Donald. Her oldest son had decided to visit, bringing flowers, which Mimi clearly appreciated. "He doesn't see me very often," Mimi said. "But we *kalked* today, and now go to each of them back more often to come each more. One crazy, you know." She laughed.

In his usual cargo shorts and untucked Oxford, Donald was seated at the foot of his mother's bed. Mimi's condition did not seem to be affecting him, at least not noticeably. Donald was so still most of the time now, it often was hard to tell what he was thinking. But Lindsay had noted that since moving to his assisted living facility, he had stepped more lightly, smiled more. "I think the social isolation that he had at my mom's house was really not good for him," she said. Debbie, Mimi's housekeeper, doubled as a part-time companion to Donald, picking him up every few days and driving him on errands, or out for walks in Woodland Park. More than occasionally, the plan would be to visit Mimi, but Donald would ultimately decide not to. "She's too bossy," he'd tell Debbie.

Today, though, he was here. And with Mimi unable to interject, Donald took over the conversation, uninhibited. He displayed a comprehensively accurate knowledge of the names of everyone in his family, including their spouses and children and the cities they lived in. It seemed he'd been paying very close attention to everything going on around him over the years. But before long, he diverged into fantasy, almost like swerving off a highway and going off-road.

"I underwrote the Academy falconry system," he said. "The mascot. I started that. I'm an architect out that way, too. I designed the cadet chapel. Our Lady of the Lords built it, but she did it at my design, to thank me for something I did."

He said Don and Mimi were not his real parents—that he was actually born five years earlier than it says on his birth certificate, and not in America but in Ireland, to a different family, also named Galvin. "My parents used the name Galvin, but they didn't come from the Galvins," he said. When his actual parents died, he said, he came to live in this family.

He referred to Mimi as his wife, and to his late father as "her husband." Don Galvin, the man who raised him, was "a saint," he said, "a neurosurgeon" who trained him in the field. But Donald chose a different path.

"I became a biological scientist, and a scientist in all fields of medicine. I have ninety thousand professions I could do, but I've done six thousand and six myself."

His favorite, he said, was "falconry."

In all of his stories, Donald seemed heavily invested in being the head of the family—the role designed for him before he got sick, and the role he cannot take on now except in his most Freudian daydreams. In these fantasies, Donald isn't just in charge, he is superhumanly potent. Donald said he sired every single member of his family, except for the ones he doesn't like: Peter, for instance, was what he calls a "swapped child." So was Matt. His siblings were his progeny, but not in a sexual way. He inseminated and created—"bred" was the word he used—his children by something he called the "American Wince," in which he just stared at someone in the right way and his seed would be spread to them.

"The way they do it is they think of their testicles, they lock in the head, and they move their eyes like this." He squinted sharply, for a split second. "It's called *wince*. The American Wince. And it gives the Dick Tracy seed—travels through the woman's eye, and mathematizes, drops down to the womb. You fill the whole body with the seed by math. And it drives in. That's how children come rightly."

When asked, Donald talked briefly about the priest he said molested him. "He was dastardly, and he was paid to hurt me," he said. He said he did not know if the priest abused anyone else, and that it happened to him just once. He seemed pretty sanguine about it now. "I got damaged and scarred and got over it. Nature heals itself."

He mentioned the medicine he must take, but that discussion spun off, too. "I'm appreciative of that," he said. "The medicine's for staph infections, for living in groups. Haldol is for living in the hallway with people. I'm a pharmacist. As an architect, I put nine thousand new pharmacies in America. So that's why I get to be a pharmacist, taking the pills. The Chinese government has challenged me to take a chance with me on that, so we can have some world conquest and pharmacy for all people. That's why I like China. I'm a neurophysiology chemist. That's what I do in my scientific field, as a scientist."

Donald smiled. So did Mimi, haplessly.

"Yeah," Donald said. "Life goes on, doesn't it?"

—⁂—

On July 13, 2017, Lindsay was in Colorado Springs for the day to help Matt. A few weeks earlier, he'd totaled his old truck, and now he needed a ride to his appointments. She took him to get his blood drawn, then to the pharmacist to pick up his clozapine, then to Matt's clinic for the proper clearance for the prescription, then back to the pharmacy. And then more errands—deliveries to two disabled friends who had relied on him for help, as long as he'd had the truck.

After dropping Matt back at his apartment, Lindsay stopped by Hidden Valley Road to see her mother. Mimi never left her bed now. Today, she was having a horrible headache. Jeff, her caregiver, had tried Tylenol and a sedative called Lorazepam, but it was getting worse.

Lindsay felt it in her stomach. This was exactly how it had started the last time, with a bad headache.

"She's having a stroke," she said.

HER MOTHER WOULDN'T let Lindsay leave her side. Every time she tried to take a break and head upstairs, Mimi would cry out as best she could through her aphasia: "Mary? Where's Mary?"

Over the phone, the hospice service told Lindsay to give Mimi more morphine than ever: 10 milligrams every hour. It took four or five hours for Mimi's pain to subside. At about 4 p.m., Mimi had a full-blown seizure. Holding on to Lindsay, shaking and out of control, she managed to say, "I'm going now, I'm going now." She lost consciousness.

Lindsay, Jeff, and Michael took turns sleeping and sitting with Mimi, administering morphine and Haldol. If they ever backed off the regimen, Mimi became highly agitated and uncomfortable. With it, her breathing was still loud but rhythmic. Through a baby monitor, they could hear Mimi's breath filling the house like a bellows. Occasionally she would stop breathing for several seconds. Each time they were sure that it was the end. Then she'd start breathing again.

Three days passed. On Sunday, Lindsay drove to Pueblo to get Peter. He brought Mimi a big bouquet of pink roses and said a Rosary for her. She got Donald from Point of the Pines and Matt from his place in Colorado Springs, and they both also had their chance to say goodbye. Mark came, and so did Richard and Renée, who cooked for everyone. John was in Idaho, planning to come out in a week's time. Margaret, on the phone from Crested Butte, said that she had made her peace with her mother already and would not make the three-hour drive to see her one more time.

IN THE EARLY hours of Monday, July 17, Lindsay administered a dose of painkillers to Mimi and went back upstairs to go to sleep. At 2 a.m., Michael heard the rhythm of Mimi's breathing change on the monitor, and he got up to check on her. He stood over his mother, watching as she inhaled and exhaled deeply, about ten times.

Finally, there was silence.

Michael woke Lindsay. Neither of them could go back to sleep. Lindsay cried and they both stayed up for a few hours, lighting candles and incense, sitting on the back deck, listening to the rain. There was something comforting about the sound of weather all around them.

The next day, the rain was still falling. Lindsay opened the front door to the house. The sky was gray, but the sun was there somewhere, giving the rain clouds above a bluish hue. Lindsay walked out into the front yard. She stood out there for a long time, arms stretched out, gazing upward as the rain covered her.

She motioned toward Michael, and he joined her. Together, they got soaked, laughing in the rain. Giddy, she tried to get Michael to dance with her, only to learn that her brother barely knew a box step. "I'm a musician, I'm always *sitting* on the stage!" Michael said.

Lindsay laughed. And when he grabbed his sister's hand, Michael froze. It looked just like his mother's hand, the way he remembered it from long ago.

DONALD
JOHN
MICHAEL
RICHARD
MARK
MATTHEW
PETER
MARGARET
LINDSAY

CHAPTER 40

On the blazing July day before his mother's funeral, Peter's room at Riverwalk—a nursing home, a few blocks away from the state mental hospital in Pueblo—had a cheap boom box blaring classic rock and a big-screen TV going full blast, both of which Peter largely ignored.

"It's *wonderful*," Peter said, looking around, Lindsay standing next to him. "I got the Bible and everything."

He showed off a photo album, filled with group shots of the Galvins. He pointed at faces and picked out names.

"Don, Jim, John, Brian, Robert, Richard, Joseph, that's me, Peter, Mary's in the chair," he said, jabbing at the photo with a shaky index finger. "They're *wonderful*. That's my dad. He was a lieutenant colonel in U.S. Air Force. He flew the falcons at the Air Force football games. The Thunderbirds were at halftime. . . . Don, Jim, John, Brian, Robert, Richard, Joseph, Mark, Matt, and that's me, Peter. Margaret, Mary"—he smiled—"that's my little girl, Mary. She's *wonderful*."

"You know, Peter, could you bring that with you?" Lindsay said.

"Yeah, yeah. I think I should, I think I should and *cooperate,* with the Bible. I love you!"

"Jeff is going to come pick you up tomorrow morning," Lindsay said. "We're just going to go have lunch today."

"Can I go with you for lunch?"

Lindsay laughed. "Sure!" This, of course, had been the plan all along.

Peter was beyond thin now—bony, his pants cinched in order to fit him. He had emerged from his room to meet his sister smiling, wearing a hockey jersey, a plaid flannel bathrobe, a ratty ski jacket, a baseball cap, heavy work boots, and winter gloves. His voice was low and gravelly, his mustache scraggly. But he still had his same puckishness, dampened only a little by exhaustion from shock therapy. Lindsay's visit happened to land on a Tuesday, and Peter had just returned from his weekly ECT at the hospital.

When Peter wasn't on a tirade, claiming his doctors were working for Satan, those same doctors found him as charming as always, even sweet. "He's the only patient I've ever gone that far out of my way where I would take him for walks," said one of his doctors, Matt Goodwin, who treated him for years when Peter lived full-time at Pueblo, and who still often administered his ECT sessions. "I would take him out for lunch." On the wards, Peter would serenade patients and doctors with his recorder, playing "Yesterday," "Let It Be," and "The Long and Winding Road." Every Christmas, he'd take out the photo of his family on the staircase at the Air Force Academy, showing everyone who was who, and talk endlessly about flying falcons with his father.

In 2015, Goodwin had petitioned the court controlling Peter's care to compel El Paso County, where the city of Pueblo is located, to make a space in one of its local assisted living facilities available to Peter. As long as he had ECT on a regular basis, Goodwin argued, Peter had no need to live inside the state mental hospital. A month later, on December 17, Peter moved to Riverwalk, which primarily serves people with Alzheimer's and dementia. Peter was by far the youngest resident there. His diagnosis: bipolar 1 and psychosis. His prescriptions: the mood stabilizer Depakote; Zyprexa, an antipsychotic; and Latuda, an antidepressant often prescribed to bipolar patients.

At Riverwalk, Peter liked to keep to a schedule. His smoke breaks had to happen at a certain time. If they didn't, he'd get agitated. "It's something for him to do out of the monotony of the day," a supervisor at Riverwalk said. "It gives him an activity." He was never violent or aggressive, though he could sometimes be loud and persistent ("You said you were going to get my cigarettes!"). He often played his recorder at a long-term care facility across the street, where the patients applauded him and asked for more. He'd play there every day if they let him.

For their lunch outing, Lindsay coaxed Peter into losing the bathrobe. It was the middle of summer. The treatment had depleted Peter. He hadn't eaten since the night before. But he was excited to leave with Lindsay. "I think I'm gonna get a big thirty-eight-ounce Coke," he said. "I want to get a cup of coffee. I like coffee. . . . I shampooed all my hair and got everything all cleaned up and put socks on and new shoes and new underwear. . . . Hey, can't we stop and get a pack of cigarettes? I want to stop and get a pack of cigarettes with a five-dollar bill."

What does he think about ECT?

Peter's expression darkened. "They knocked me out. They knocked me out cold with oxygen."

How does he feel afterward?

"I just cooperate fully and do everything that they say."

On the way out, Peter stopped in the lobby, pulled out his recorder, and performed a Christmastime favorite—"Angels We Have Heard on High"—before walking stiffly out the door.

"I want to go get a burger!" he said in the backseat of Lindsay's SUV. He flashed cash from his wallet. "I got all the money. Twenty-five dollars, right here."

"That's all right, I've got it," Lindsay said.

"Okay, I'll cooperate fully."

"So it's going to be a big crowd tomorrow, Peter," Lindsay said.

"Yeah, it will be."

"Do you have something nice you can wear?"

"Yep."

"All of Mimi's grandkids and great-grandkids will be there."

"I'm going to go smoke! I wish I could have gotten cigarettes."

"After lunch we can go get some cigarettes."

They pulled up at a pub in downtown Pueblo, where Peter ordered a large Coke and a burger with fries and ketchup, tearing through the fries first. Some Riverwalk employees noticed him from across the room and walked over to say hello, smiling and asking how Peter was feeling today.

"So who were they?" Lindsay asked, once they returned to their table.

"I don't know," Peter said.

"Were they from the hospital?"

Peter did not answer.

"Are you feeling okay?"

"No. I'm sick of everything that I went through. I want to get a pack of cigarettes and *cooperate*. I'll go buy it myself and cooperate with you in full to do everything you want me to. Just don't smoke 'em. I'll smoke 'em myself. . . . I can't eat this ketchup with cheese. I think I have an upset stomach. The ketchup makes me feel funny. . . . I'm cooperating *full*. I want to cooperate—do anything for you that I can."

AFTER LUNCH, LINDSAY pulled up to a store and let Peter out to get cigarettes by himself, giving her a moment to speak openly about his condition. "Dr. Freedman explained it to me," Lindsay said. "Years and years of overmedicating. That's why they do these ECTs, because the medications really don't work for him." This was a version of the same problem all her sick brothers had. The less consistently you take the medication, the worse off you were—the more psychotic breaks you have, the more far gone you become. It was a painful catch-22 to witness a loved one experience: Not taking the drugs makes them more sick, and then taking them, in some cases, makes them sicker. A different kind of sick, she agreed, but sick nevertheless.

"He said eventually the medications will have no more impact," Lindsay said. "And it's really the ECTs that have caused the majority of the memory loss. This is more disorganized thinking. Not able to answer questions. And the mantra—*I cooperate fully*—is constant."

That saying, so specific, must have some meaning to Peter. All those years of parents and doctors telling him he was not cooperating, Lindsay said—maybe they've made a mark on him.

Peter hopped back in the car, smiling. "God, that's fast in there. I've got a whole pack. Can I light one up in here?"

"No!" Lindsay said cheerfully.

"Okay," Peter said, then muttered: "I'll cooperate fully." A moment later, he brightened again. "I have a whole pack of Marlboros. You people are *wonderful*."

—ᴡ—

Lindsay's next stop that day, Matt's home at the Citadel Apartments in Colorado Springs, was a small, no-frills place paid for with a Section 8 housing voucher. Never one to focus on his personal hygiene, Matt nevertheless maintained his home like the tidiest of hoarders, his towering piles of stuff always neat and organized. "I bet he's got a fortune in collectible vinyl," said Lindsay as she pulled into the parking lot.

Matt's most prized collection was his stack of Clint Eastwood movies—DVDs and VHS tapes. Most of the time, when Matt was on the phone with his family, *A Fistful of Dollars* or *The Good, the Bad and the Ugly* could be heard blasting in the background. "I told him that Clint Eastwood is Republican," Lindsay said, smiling. "That was very disappointing to him." But he still watched all the movies.

Visits and phone calls with Matt were never predictable. Sometimes he'd rage about being labeled mentally ill, about his mother putting him on medication, about the millions of dollars he said the government owed him for building all of the roads and bridges in the state of Colorado, about how the mental health profession had killed his father and two of his brothers, Jim and Joe. "They might as well kill me!" he'd moan—he had nothing left to live for. But today, the day before his mother's funeral, Matt was in a decent mood—not delusional, just glum, and, as usual, a little caustic. He'd been watching *Hang 'Em High* when Lindsay showed up. In jeans and a leather biker vest, he was a little imposing, tall and stout with unruly long hair, a scraggly beard, and the same deep-set eyes as his brother Donald. Lindsay's kids, whenever they saw him, always remarked on how much he looked like Hagrid. Even his voice was a low, mumbly growl.

"Well, my shoulder couldn't get any worse than what it is," Matt said, sinking into the backseat of the SUV.

"You've got a doctor's appointment, though!" Lindsay said, triumphantly. Seeing doctors had never appealed to Matt. For years, Lindsay had been trying to get him to get his teeth fixed, but he thought the dentist would implant something in his head.

"I've got an appointment over at Park View on the tenth of August," he said, and then he started running through other old business, concerned about tying up loose ends from his accident with the truck—the one Lindsay had been helping him with just before Mimi's death. Matt actually had been in the middle of a good deed when the crash happened. He was helping his friend Brody, a Vietnam vet who is a paraplegic, get to Denver to get a new bag for his catheter. They were on their way back during rush hour on a Friday night when Matt saw a car stopped in the center lane and slammed his brakes. He missed that car, but then the two cars behind him smashed into him, one after the other.

"They sent me a letter from the impound lot, saying it cost eight hundred and fifty dollars?"

"I know," said Lindsay. She had spent hours on the phone with the police and the courts and the insurance company, sending copies of the power of attorney document that Matt had signed to show that she could handle everything on his behalf. "If they call you or anything, or write another letter, give it to me."

"I just want to sort that out."

"We will. It's gonna take a long time, though, Matt. The courts, they haven't even assigned a permanent case number to it yet."

Lindsay tried to bring up tomorrow's funeral, just as she had with Peter. Matt also didn't pick up on that. Instead, over a sandwich at a nearby sub shop, he ran through a litany of his many injuries and wounds. "I had six separate teeth surgeries. And I had a blot clot removed from my brain in 1979, I was twelve and a half."

"I was at that hockey game," Lindsay said.

"It was at the Air Force Academy," he said. "It was the league championship. We beat Mitchell. They had twenty-two players, two goalies, and a coach. We had eleven guys. You know what you say about hockey? Go puck yourself."

Lindsay smiled. She was used to Matt's jokes. Most were dirtier.

"Our team went to state," he said. "But I couldn't play because I broke my face. This guy picked me up under my butt and threw me into the boards."

"I remember!" Lindsay said. "I sat next to you in the backseat of the car and your eyeball was hanging out of your face."

He showed Lindsay a scar on the side of his face.

"I got a hundred and fifty-seven stitches," Matt said, launching into his usual exaggerated version of the story. "I flatlined and they used the shockers. You know that *ER* show, with the shockers? They hit me ten times, and I flatlined for seven and a half minutes, and they said do it one more time. The eleventh time they hit me, they got a pulse, and I woke up two and a half weeks later."

He reminisced a little about his college days at Loretto Heights—girls in the dorms, Frisbee in the hallways, all the hockey players he knew there. He remembered dropping out after a year and working at the bowling alley and having a newspaper route and living with his brother Joe for a while.

"When Joe died, me and Mark and Mike went out there and divided his stuff between the three of us," he said somberly. "I got his TV."

The subject of Joe propelled him into more difficult territory. "Donald just made my life a nightmare," Matt said. "He took his anger out on the whole family. He smacked me across the floor." The more he talked about his childhood, the more he descended into self-pity. It was never far from Lindsay's mind that Matt—who had once been the coach of her soccer team; whom she once wrote an essay about, calling him her hero—really was a victim, just as she was.

"Donald, Brian, Jim all abused me," Matt said—though, given this was Matt talking, there could be no way of knowing how true that was. "So I left the family for like eight or ten years. And I came back, and Jim had a heart attack, over there at Main Street. And Joe had a heart attack. And my dad died. And then my mom died. And I lost my family. And there's nothing I can do about it."

"I'm here," Lindsay said.

Her brother glanced at her. "It's good to see someone still here."

—◦◦◦—

That night, Mimi's house on Hidden Valley Road received a host of Galvins who had come to town for her funeral. Michael drove in from Manitou Springs with his wife, Becky, and one of his daughters; he was still unpacking the experience of taking care of Mimi as she left this world. "I told Mary that taking care of somebody like that, it's really a privilege," Michael said. "Because if you *had* to do it, you would. But because there's enough money, most of us don't have to."

"Hey, sunshine!" John said, spotting Michael.

John, the music teacher, now retired, had come down from Idaho with Nancy—their first time back at the house since his mother's ninetieth birthday, three years earlier.

Michael brightened. "Hi, there he is!" The two brothers hugged. "I think you shrunk a couple inches, buddy."

"Well, maybe a little bit," John said.

"No, I'm sure you did," Michael said. "You were always taller than me, weren't you?"

"Well, yeah," said John. He'd fallen off a ladder two years earlier and endured a long, painful recovery. "Three back surgeries, four knee surgeries, three ankle surgeries. I was one step above an invalid for the last two years."

"Hey, I got some ladder work if you want to do it," Michael said with a smile.

John and Nancy had come to town in their RV, a retirement splurge. Entering their golden years in Boise, they had some creature comforts now: an antique piano they meticulously restored themselves, a koi pond in the backyard, and a small arbor where they grew grapes for wine they made in small batches and labeled. They used the RV to travel the country, making trips to Colorado somewhat more feasible. But they had built a life apart from the Galvins, in part by design and in part, they said, by necessity. "Margaret and Mary have probably taken the brunt of all of it as far as taking care of those who are mentally ill and seeing to their needs," John said. "And they have the money to do that."

Now that he was here, John was already feeling a little put out. He had rehearsed a piano piece for his mother's service, only to learn he would not be able to play it. Lindsay had planned a gathering outside, in a meadow. He'd wished they'd put together something more formal for his mother—even though rationally he understood that he had no real right to feel that way, given that the timing of the funeral was arranged around his previously scheduled visit to Colorado Springs. Still, it was unsettling, not the closure he wanted. John found himself in a narrative that was unfamiliar to him, one he could not control. This is the way it works, a lot of the time. If you've left town, like John, you can hold on to your truth. To come home is to run the risk of being contradicted. Even the people who leave, like John, can feel almost rejected.

John decided a long time ago to live his own life the best he could, but he never saw a role for himself in caring for his brothers. "I try to see Matthew and Peter if they're available when I go down there, maybe once a year," he said. "But my oldest brother, Donald, well, you couldn't have a conversation, basically."

MATT DECIDED NOT to come to dinner; he had fared well enough during lunch earlier that day with Lindsay, but seeing everyone at Hidden Valley Road seemed difficult for him. Peter was not invited to dinner; for him to mix and mingle with family the night before the funeral seemed like too much—too exhausting for him and everyone around him. But the next day, both Matt and Peter would be at the funeral, Matt skulking in the background, sweating uncomfortably, and Peter beaming in front of everyone, the closing act of the service, playing "My Favorite Things" on his recorder to a round of applause, and then, for an encore, reciting a rambling, customized version of the Nicene Creed: "I believe the one God the Father Almighty maker of heaven and earth. . . ."

Mark Galvin came in from Denver for the funeral-eve dinner—the eighth son, the onetime hockey star and chess prodigy, and now the youngest Galvin brother who was not mentally ill. Bald with a goatee and a wide frame, Mark resembled no one else in the family, except perhaps in the way he talked. He and John and Michael all spoke high-mindedly about politics and music and chess—cultured in the fashion their mother had always hoped. He had retired from managing the university bookstore—a state job with a pension that he'd started collecting. In his retirement, Mark had turned his car into a private taxi service, doing regular business at two of the fanciest hotels in Boulder, the St. Julien and Boulderado. This new career had caused him to cross paths with some people that Mimi would have loved to hear about, like the artistic director of the Boulder Philharmonic, who hired Mark to take her guest artists to and from the airport. "I've got tickets to Vivaldi in January," Mark said. "I'm driving Simone Dinnerstein"—the world-class pianist—"back to the airport from Boulder, after getting free tickets in exchange."

Mark had felt alone in his family for decades now, the other hockey brothers dead or sick. Some days his entire childhood seemed like a blank to him—an impulse to move on, perhaps, or to stop hurting. A few of the more vivid memories, however, hadn't faded.

Mark had an excellent recall of the massive blowup between Donald and Jim on Thanksgiving, forty-five years earlier—and Donald picking up the dining room table and throwing it at Jim. "A madhouse," Mark said, shaking his head.

OF THE WELL siblings, only Richard and Margaret didn't come to the dinner on Hidden Valley Road. Richard seemed to be avoiding a confrontation. He had recently launched an email salvo against Lindsay over the subject of Mimi's will, arguing that Lindsay should not be the executor, only to get pushback from all the other well siblings, who came to Lindsay's defense.

In Lindsay's opinion, Richard was just upset not to be included in the will. Lindsay said that Mimi had made the decision to leave him out only because Richard had already accepted money from Don and Mimi several years earlier, to help him through a rough patch. "My father couldn't stand Richard," she said. She could not deny, though, that Mimi had thought the world of Richard, laughing and gossiping with him whenever he visited. "She would play us against each other to get what she wanted," Lindsay said. "That's a trait I have to work very hard not to have."

When Michael used to watch Richard cozy up to Mimi this way, he'd almost laugh. "He wants so much to be like his father and feel on top of the world," he said. "I think he tries too hard."

To hear Richard tell it—over lunch one day, a few weeks later—he clashed with Lindsay because it seemed to him that all she ever wanted to discuss was the sick brothers. "I got so upset. I said, 'Mary, I want one dinner to talk about the moon, the stars and the skies without talking about mental illness.' It just became so depressing for me."

Richard seemed to take more after his mother than his father, determined to speak about pleasant subjects only, like his trips to Pebble Beach and Cabo, and his business deals in Dubai. Like Mimi, Richard also was convinced of the value of having a pedigree, being raised from good stock. This much was clear when he told stories about his father that were unlike any that anyone else in the family told. In Richard's version of his father's life, Don Galvin wasn't the second-in-command of the USS *Juneau*—he was the captain. Don Galvin wasn't just a briefing officer at Ent Air Force Base—he had a personal relationship with President Eisenhower. Don Galvin wasn't just the first executive director of the Federation of Rocky Mountain States—he founded it. Don Galvin didn't get his Father of the Year award from the Knute Rockne Club—the award came directly from President Nixon. Don Galvin wasn't just the president of Colorado Springs' local ornithological group—he "brought Audubon to the West."

And Don Galvin wasn't just a communications officer at NORAD. "Dad was in OSS," Richard said, "which became the CIA."

Richard would talk at length about covert missions his father took to Iceland, Ecuador, and Panama, all while using his jobs at the Academy and NORAD as covers. All this, Richard said, he'd gleaned from conversations with his mother. "She just said there were things that he could never say," he said.

The idea that Don Galvin was a spy is unsubstantiated by any available information from any military branch or intelligence agency. And yet this romantic view of his father was helpful to Richard. At the very least, it was preferable, for instance, to the story of a father whose military career stalled out—perhaps because he'd harbored the liberal political views of an academic, not the hawkish view of a military officer—and who gritted his teeth after being demoted to service as a glorified PR man.

Rather than think of Don Galvin that way, Richard adopted a convenient self-delusion. Not the sort of delusion that fits a DSM criterion. But we all have stories we tell ourselves.

MARGARET HAD TOLD Lindsay that she didn't want to spend the night at the house—that she'd rather come in for the funeral the next morning with Wylie and her two girls. Once again, Lindsay felt abandoned. She was not sure what to do with that feeling. Most of the evening, she didn't discuss it—until, in the kitchen, John turned to Lindsay.

"So. Margaret's not here."

"Yeah, whatever," Lindsay said.

"What's the problem?"

Lindsay took a few seconds, not sure how measured to make her response.

"I think it's Margaret's overwhelming guilt," she said finally, "at not having lifted a fucking finger for, like, *ever*."

"Yeah, she's into her own thing," John said, treading lightly.

"She is *into her own thing*," Lindsay said, and her smile widened. "Actually, there you go! That is the explanation."

LINDSAY WALKED OUTSIDE to the patio and hugged Michael and Mark. There was talk of who had RSVP'd for the funeral and if the clear weather would hold long enough before an expected rainstorm. Then the reminiscing started—the epic road trip the family took across the country for the 1964 World's Fair in New York; the luggage flying off the roof when Dad misjudged the clearance of an A&W restaurant drive-through; all the luggage coming into the car, jammed in with the kids and the birds.

"Didn't he drive off the road in Kentucky in another rainstorm?" Mark asked.

"Yeah," said John. "And in the rainstorm a rock hit the truck, the bus. And then he had to take it to New Paltz, New York, to a repairer-dealer. He dropped the screw into the rotor. The mechanic found the screw in the rotor."

"I remember the rainstorm," Michael said, "but I don't remember the other stuff."

"You don't remember the rock hitting the van?" said Mark. They all laughed.

"Who the hell keeps falcons?" Lindsay said. "Every time I tell people, they're like 'What?'"

"I tell people stories in the cab all the time," Mark said.

John turned to Lindsay, suddenly serious, thinking about the funeral.

"What's plan B if it rains?"

"Umbrellas," Lindsay said. "If it rains, John, you can play at the restaurant."

"The keyboard's electronic," John said. "It's just not the same."

Lindsay smiled and motioned over to the piano that Mimi had still kept at the house. "I'll try to convince them to take the piano up from the basement and out to the field."

There was more laughter.

DONALD WAS ALONE in the living room, away from the others, smiling politely at anyone who smiled at him. Today happened to be his seventy-second birthday, and Lindsay had asked Debbie to get him a cake as a surprise. But he kept to himself, mostly silent, until he was asked if he'd had a chance to say goodbye to his mother.

"Yes, when she first left," Donald said. "She said, 'Thanks.' I said, 'Thanks,' back to her. I just thanked her for being there."

Will he miss her?

"No," Donald said. "She's bred. She's out of harm. I mean, she's at sea right now, as a triplet."

His mother is a triplet?

"I bred her as a triplet, at sea right now."

As a human being, or as a fish?

Donald scowled, finding the question ridiculous. "As a human."

But she's at sea?

"Yeah," Donald said. "They live with an octopus."

A human lives with an octopus?

"Yes. Octopuses have the ability to make man. To make many humans, all animals. When the flood comes, then they keep them alive in the water sometimes."

And Mimi is there, as a triplet?

"Yeah. She's a little one right now. A little baby. She's out there, maybe five months old today."

Would you like that to happen to you when you die?

"Oh, I wouldn't mind," Donald said.

JUST BEFORE IT was time for Donald to return to Point of the Pines, they brought out the cake: chocolate with cut-up chunks of a Snickers bar on top. Donald had been so quiet all evening that he was almost not there, a shadow. But he seemed pleased by the attention now, smiling softly, his lips never parting.

Debbie lit the candles and brought the cake out to the patio where everyone was sitting—the same patio where they'd once kept Frederica and Atholl, and where Matt's head slammed to the ground in a battle with Joe. As everyone sang "Happy Birthday," Donald—the oldest person in the room now, the paterfamilias—stood over the candles and broke out into a wider smile. Then he crossed his arms across his chest and closed his eyes, as if he were making a wish.

Part Three

DONALD
JOHN
MICHAEL
RICHARD
MARK
MATTHEW
PETER
MARGARET
LINDSAY

CHAPTER 41

Lindsay had left Hidden Valley Road when she was thirteen, determined never to come home. She had moved from Boulder to Vail and then to Telluride, keeping her distance. But now, with Mimi gone, she was back there more often than she had been in years, seeing Donald, checking in on Matt, driving farther out to see Peter, and prepping the house for sale. As Lindsay drove the streets of Colorado Springs, memories revealed themselves to her—like the cottages west of the city, not far from where she had once hidden out with Kathy, when Jim got violent. "I drive by that all the time now," she said.

She still felt like the youngest—like everything the family went through flowed down to her. Part of her will always want vindication—and she may always feel a little abandoned, a little insecure, tiptoeing along a knife's edge. This might explain why she was working more than ever now, in addition to assuming the responsibility for her sick brothers' medical care. Some days, she recognized the blessings of being detail-oriented, hyper-vigilant. "Louise joked in therapy—it's only a red flag when it starts to create conflict in your life, but otherwise it's a truly healthy coping mechanism for you to organize your sock drawer." She laughed. "I'm very tidy."

Her decision to do all this—to stay, and not drop everything—
was as much of a mystery to her as it always had been.

"In all that therapeutic work," she said, "the therapists I've had
have been like, 'Holy shit, you've got to be kidding me. You *survived*
that?' But what was the alternative? Succumbing to it? What would
that look like? Be a heroin addict? I don't know. As a child and for
years into my young adulthood, I deeply wished that my brothers
with mental illness would just die. But that was a gut-wrenching
wish—it tore at me."

A FEW MONTHS after the funeral, the house at Hidden Valley
Road went on the market. In the summer of 2018, the eventual buyer
emailed a note to the broker.

> Good Morning Galvin Family,
> Thank you for allowing my husband & I the pleasure of
> viewing your family's home last night—it is truly incredible.
> Walking through the home we could clearly see the care &
> the loving memories that went into this house and imme-
> diately wanted to continue its story. We hope that you will
> thoughtfully consider our offer as we would love to build our
> family there.
> Thank you & we hope you have a wonderful day!

During one of her visits to Colorado Springs, Lindsay took a side
trip to the state mental hospital in Pueblo to unearth what still survived of her brothers' old medical records. Maybe she should have been
prepared for a few more family secrets to be revealed. It was in a sublevel of the hospital's main building, sifting through those papers—
two shopping carts full of overstuffed accordion folders, pages poking
out in every direction—that she first learned about Donald's attempt
to kill himself and his wife, Jean, with cyanide and acid. For all those
years, Mimi had said merely that Donald became ill because his
wife left him. The truth was something quite different, an attempted
murder-suicide, not unlike Brian and Noni, three years later.

Lindsay also saw the medical report from Colorado State in which
Donald talked about trying to commit suicide when he was twelve
years old. This, too, was something no one in her generation ever

knew. If Mimi had known, she'd never discussed it; again, it seemed easier, perhaps, for her to decide that it all went wrong for Donald after he left home, and not while he was in his mother's care.

When Margaret learned this, she felt bamboozled all over again. "*I* had no idea Donald tried to kill his wife," she said. "That also explains so much to me. I was never satisfied with the answer I was given—which was vague and only that he was getting sick." Until the day she died, Mimi had preserved some of the illusion—maintaining the "before" picture, until there was nothing left to protect. Margaret couldn't help but wonder what might have changed if her parents had been more forthcoming about Donald, if everyone had known what he'd tried to do with Jean. Would there have been more sensitivity about Brian's state of mind? If her parents had been just a shade less secretive, could someone have prevented Brian from doing what he did? Would Lorelei Smith still be alive today?

The secrecy felt like an insult to Margaret—another rejection. "I was fed a line of bullshit from my parents. I think they must have wanted me to believe Donald was better than he was."

At Pueblo, Lindsay found paperwork on all of their brothers, as well as a file about their father that offered yet another surprise. For several years before Don died, Lindsay learned, he'd been traveling to Pueblo on a regular basis for ECT sessions. The stated reason was depression he'd been experiencing since the early 1990s, after multiple occurrences of cancer and the death of one of his brothers. But of course this new information only brought on more questions. Was their father having ECT because of a clinical depression that was genetic, tied to schizophrenia? Was this the same condition that had hit him in Canada in 1955, as Mimi had thought? Or was Don caught up in an entirely new depression at the end of his life, because who wouldn't be, in his situation—with one of his sons dead in a murder-suicide, another five hopelessly delusional, one a compulsive child molester? After so little about his life had turned out even remotely the way he'd wanted?

Mimi had to have known about Don's ECT sessions. She'd gone there with him, and no doubt driven him home afterward, as often as once a month for years on end. She'd kept this secret, too. To be a member of the Galvin family is to never stop tripping on land mines of family history, buried in odd places, stashed away out of shame.

Lindsay didn't know how to react to this one, except to muse yet again about the damage caused by that secrecy, and to try to live her own life differently. Maybe, she thought, her family's story was not just about the secrets, not just about a disease—but about how all of that experience, with the help of Drs. Freedman and DeLisi, might make life better for others.

Was it worth it to them? Not really. But maybe there was something for her to hold on to now, with Robert Freedman's choline trials and Lynn DeLisi's SHANK2 revelation—a sense that their sacrifice may make it better for future generations. Isn't that how science works—how history works?

CHAPTER 42

One night a few years before Mimi got sick, Margaret woke up crying from a dream that was too much for her to bear.

In the dream, she and her sister were in Vail after a day of skiing. Lindsay didn't say where they were going—and the knowledge that her little sister knew and Margaret did not is, perhaps, a telling detail in its own right—but soon Margaret realized they were heading toward the condominium owned by Sam and Nancy Gary. When they arrived, the door was unlocked.

Lindsay walked through, and Margaret followed her. They were alone. The place was not in the best shape. Lindsay said that Sam's children use it now. That got Margaret thinking of all the Gary family members she once knew and had not seen in years. Sure enough, Nancy and Sam came through the door, along with their children and their friends. Clearly, they were having some sort of party to celebrate something.

Margaret felt awkward. She did not know why she was there. Only when she noticed her sister using a measuring tape to gauge the size of the room did she understand. They had been asked to help plan a party for Sam and Nancy.

It was too late now to set up. More guests were coming through the door, filing along a wooden walkway into the living room. Margaret saw Sam's secretary, the Garys' drivers, their cooks, their housekeepers, even the tennis instructor who came to Montana to give Margaret and the others lessons by the lake house. They all were older now, but Margaret recognized them just the same.

She was uneasy, convinced she did not belong there. Then one of the family's tutors came up to her and smiled. "I don't know why I've been gone so long," Margaret told him. "You all are such great people." The tutor replied, "Well, we've got to get you into our family history."

Margaret felt better, but the feeling didn't last. She overheard other guests mentioning other parties she hadn't been invited to. Suddenly it all came back to her—the one-upmanship of the Denver social scene, how she never fit in, and how the only reason she ever came into contact with it was because of the breakdown of her own family. Everything came back to that deep well of rejection—of pain. Then came the tears.

WHEN YOU DON'T find a sense of love and belonging where you are, you go searching for it somewhere else. In Margaret's case, and perhaps Lindsay's too, the first stop in that search had, tragically, been Jim's house—a place away from home, with a family member who paid attention to her. For Margaret, the Garys' home and the Kent Denver School represented more chances to belong somewhere—problematic, too, in their way.

Then came Margaret's Deadhead years, traveling with a tribe of like-minded nomads, and her brief first marriage. Looking back, she felt lucky to have survived. *Did I really marry a guy who dealt drugs when I was twenty?* she wrote in her diary.

And then finally her decision to settle down with Wylie and have a family of her own. "I like to call him a safe harbor," she said.

In the years when she and Wylie had their daughters and Margaret became a full-time mom, she grew preoccupied with maintaining some sense of emotional equilibrium. "You're the *feeler* of the family," Mimi often told Margaret, and on this point, at least, Margaret and her mother agreed. In therapy, Margaret had said that Brian's death had been the pivotal moment of her childhood, as searing, even, as the abuse she experienced; she was eleven at the time, old enough to see the toll it took on everyone. But the trauma she dwelled on most often was abandonment—not just being sent away to the Garys, but being neglected before then, too, in favor of so many other siblings. "The kids who don't get the attention are the ones who often need it most," Margaret said. "At least that was my experience."

Margaret thought often about something her mother always said of her and her sister: "The roses after all the thorns." She and Lindsay were the roses, and all ten of her boys were the thorns. What most people saw as tender struck Margaret as ugly and passive-aggressive. What must the boys have felt, growing up hearing their mother say that? And how could the girls be secure, hearing praise for them in the same breath as such dismissive scorn?

As one of those two roses, Margaret never felt she had a shot at her mother's love. If Mimi really loved her, she never would have sent her packing at the age of thirteen. Sometimes Margaret felt that her time with the Garys permanently separated her from her mother—that she had never gotten over that rejection and had spent the rest of her life trying to protect herself from being hurt that badly again. *I've already been cast aside as a throwaway, a cast-off,* Margaret once wrote in her diary. As time went on, she felt more of a right than ever to create distance between herself and everyone else. *I want the closeness of a normal family, but frankly my family of origin is not normal.*

To Margaret, her sister and her mother seemed like two peas in a pod. Mimi gave Lindsay furniture from her house and even sewed clothes for her, and Lindsay seemed to show no ambivalence in the slightest about taking care of Mimi in return. Margaret resented them both sometimes, though she needed them both, too.

ONE OF MARGARET'S most vivid memories from just before she was taken from Hidden Valley Road—those months after Brian died, when she watched her father and her brothers falling apart all around her—was her mother staying up late, long after the children were in bed, to draw and paint—birds and mushrooms, mostly. When Margaret thought about that later, she was beyond confused. How could Mimi still be puttering around the house, watching for the fox and the family of deer that ambled by the backyard, reporting on the dramatic loss of birds at the bird feeder? This was the same woman she'd just seen wailing with grief over Brian. What did her mother have inside that Margaret didn't? Was it strength, or denial, or something she couldn't understand? Only later did she arrive at the idea that the natural world Mimi had fallen in love with in Colorado offered her some small measure of solace, a refuge from everything else that was happening.

Once Margaret, in her adult years, finally worked up the bravery to start painting, her subject, more often than not, was the very thing she had spent a lifetime trying to avoid: her family. She painted flowers that her mother loved, with a stirring realism. She made one painting about the Garys, called *Gray Ease;* another called *Sophisticated,* about her own journey, learning to be vulnerable; and another called *Compartmentalizing the Grief.* She veered into abstraction in a striking series of twelve paintings based on the twelve Galvin children. *Donald* is red and white; *Jim* is a spectral black and white; *John, Brian, Michael,* and *Richard* are variations on greenish yellow; *Joseph* is yellow with red seeping through; *Mark, Matthew,* and *Peter* are all studies in red, with only Peter's including flashes of blue.

Mary is a cross-hatch of thick streaks of soft pink, inflected here and there with black. And Margaret's self-portrait is similar to her sister's, only with less pink and more vivid rust-colored flecks.

When, a few years before Mimi died, Margaret helped relocate Peter to his assisted living facility, that inspired another piece, *Moving Peter,* that seemed like a step forward for her—complex and layered and full of the feelings she found so hard to process any other way. "It just became this emotional outpouring," Margaret said.

This was the painting Nancy Gary bought, snapping it up before an old classmate of Margaret's from the Kent Denver School had the chance to buy it.

CHAPTER 43

Our culture looks at diseases as problems to solve. We imagine every ailment to be like polio: hopelessly incurable, until a miracle drug comes along that can wipe it off the face of the earth. That model, of course, only works some of the time. Too often, scientists get lost in their own silos, convinced their theory works to the exclusion of everyone else's. Whether it's the Freudians and the Kraepelinians or the family dynamics specialists and the geneticists, the unwillingness to collaborate leaves everyone vulnerable to confirmation bias—tunnel vision. The schizophrenia researcher Rue L. Cromwell described this dilemma in the 1970s: "Like riding the merry-go-round, one chooses his horse. One can make believe his horse leads the rest. Then when a particular ride is finished, one must step off only to observe that the horse has really gone nowhere. Yet, it has been a thrilling experience. There may even be the yen to go again."

But there is another model for progress—the opposite of the polio model—one in which solutions are not the same as breakthroughs. Progress comes gradually, often painfully, in fits and starts, and only after many people spend their entire careers failing and quarreling and, finally, reconciling. Sooner or later, some ideas fall away as others take hold. And, perhaps only in hindsight, we can see how far we've come, and decide on a path forward.

What would progress look like for schizophrenia? If the Galvin boys had been born a half century later or more—growing up today, let's say, and not in the 1950s or 1960s—would their treatment be any different now? In some respects, little has changed. The market for new schizophrenia drugs remains sluggish. Antipsychotic drugs require expensive and risky testing, even in the early trials, where rats are no substitute for humans. And the same nature-nurture squabbles over the source of the illness have continued, if at a more granular level. Where the conversation once was about Freud, now it's about epigenetics—latent genes, activated by environmental triggers. Researchers now argue about what might be playing the part of a trigger—something ingested, like marijuana, or infectious, like bacteria? Researchers have come up with a variety of other suspects—head injuries, autoimmune diseases, brain-inflammation disorders, parasitic microbes—all of which have their adherents and detractors. Everyone still picks their horse on the merry-go-round, and very few are willing to stop taking the ride.

There are, however, more subtle changes—as if the atmosphere around the disease has changed a little, charged with a new sense of tolerance. Anti-psychiatry, in its latest incarnation, has become a movement concerned with legitimizing and normalizing the concept of hallucinations—a Hearing Voices Movement, not unlike the movements to legitimize deafness and blindness not as disabilities but as differences. Neurodiversity—a term used more often for other conditions, like autism—is a concept that was never considered when treating any Galvin brother decades ago. There is a robust anti-medication movement now—activists armed with studies showing that many schizophrenia patients experience favorable long-term outcomes without prescription drugs. This movement has support from many therapists unhappy with the notion of psychiatry as a pill mill, and nostalgic for the gilded age of psychotherapy, when a doctor could spend more than just a few minutes with a patient before sending them off with prescriptions.

If there is a significant change, it's that more people are acknowledging the elusive quality of schizophrenia diagnoses, aware that there is no one-size-fits-all definition. Each passing year brings more evidence that psychosis exists on a spectrum, with new genetic studies showing overlap between schizophrenia and bipolar disorder, and bipolar disorder and autism. The most recent research suggests that a surprising number of us may be at least a little bit mentally ill: One meta-analysis, published in 2013, found that 7.2 percent of the general population has experienced hallucinations or delusions; another study in 2015 put the figure at 5.8 percent. A third of the people counted in the latter study only had one episode, while others had more persistent symptoms. Results like this suggest, at the very least, that the medical response to aberrant behavior should be more discerning about who needs traditional treatment and who might benefit from watching and waiting. The stakes in such decisions are high: Researchers have the evidence now to confirm that each successive psychotic break causes more permanent damage to a brain, a further loss of gray matter necessary for processing information.

The grave dilemma of neuroleptic drugs, sadly, remains the same: Medication, taken regularly, can stave off further breakdowns (while risking long-term side effects), but there is also ample evidence that patients who remain on drug regimens relapse as often as those who don't. While the surviving Galvin brothers are as dependent on neuroleptics as ever, the biggest change for those who come after them could be that medication and therapy are not an either/or choice. Even the most traditionally trained schizophrenia researchers are pushing what Jeffrey Lieberman, the chief psychiatrist of Columbia University Medical Center of the New York-Presbyterian Hospital, calls an "early detection and intervention model of care." A relatively new wave of research supports the effectiveness of so-called "soft interventions": a mixture of talk therapy and family support, designed to keep the amount of medication to a minimum. For decades, countries like Australia and Scandinavia have used this more holistic approach and reported success. (You could argue that Michael Galvin found his soft intervention inside the Rock Tumbler on the Farm, his commune in Tennessee—assuming he was ever really at risk to begin with.) The challenge is being able to tell who can be successfully treated by neuroleptic drugs, who might not be helped much at all by those drugs, and who, in the long run, might suffer from the drugs as much as the disease.

For more researchers, the watchword is prevention—the challenge of accurately diagnosing people at risk of developing schizophrenia *before* their first psychotic break. Lieberman at Columbia is developing new techniques to measure the function of the hippocampus. In time, new drugs could stave off the onset of schizophrenia—just like the drugs being developed now that might fend off the symptoms of Alzheimer's disease. And then there's choline. In Denver, Robert Freedman is following up on his first long-term choline study with a new trial—with support from Sam and Nancy Gary, among others—tracking children from the moment their expectant mothers start taking the choline supplements, up through the time their children reach post-adolescence, the prime years for the onset of schizophrenia. As he suggested at his award ceremony in New York, Freedman will undoubtedly not be alive when the results come in. Neither will the Garys or many of the other donors. "They're a bunch of builders, developers—oil barons like Nancy," Freedman said. "They said, 'Oh, yeah, let's go all in. That's how we run our businesses.'" If it doesn't work out somewhere along the way, they told Freedman they'd all have one dinner and say it was a nice ride.

Freedman also started a collaboration with the Lieber Institute for Brain Development at Johns Hopkins University—cofounded by Daniel Weinberger, the developmental hypothesis author from NIMH—to focus on fetal health from a new angle: studying whether the risk of schizophrenia is linked to the condition of an expectant mother's placenta. With Freedman, Weinberger has begun investigating whether choline might play a role in improving placenta health. Both researchers hope to eliminate a large number of potential schizophrenia cases in one fell swoop, before the patients are even born.

For Freedman, prevention is more than just good medicine; it's common sense. Billions of dollars are spent each year on developing drugs to treat the symptoms of mental illness *after* it already manifests. What if some of that money were spent on prevention, not just in the womb but in childhood? Think of all the young people who develop mental illness out of sight of anyone who can help them. What if some of those breakdowns—even suicides—could be prevented, by shoring up the mind's vulnerability before things get worse? "The National Institute of Mental Health spends only $4.3 million on fetal prevention research, all of it for studies in mice, from its yearly

$1.4 billion budget," Freedman noted recently. "Yet half of young school shooters have symptoms of developing schizophrenia."

There is no way of knowing how life might have been different for the Galvin brothers if the culture of mental illness had been less rigid, less inclined to cut people off from mainstream society, more proactive about intervening when warning signs first appeared. But there is, perhaps, reason to hope that for people like the Galvins born fifty years from now, things could be different, even transformed.

—⁓—

"I believe the trend is coming back to families," Lynn DeLisi said, over coffee a short drive from her home in Massachusetts. In 2016, the same year as her SHANK2 study, she published a paper in *Molecular Neuropsychiatry* arguing that researching families with schizophrenia was more important now than ever. For the first time in a long time, she is not the only scientist saying this.

"I think the families have enormous importance," said Daniel Weinberger. Once upon a time, when he worked alongside DeLisi at NIMH, Weinberger had been skeptical of studying families, all but dismissive of her approach. Now, like DeLisi, he sees the value of using families as workshops—or test kitchens—for theories that emerge from a GWAS. "Ultimately, families will be critical to translate the genetics into how individual people get sick." Weinberger recognizes how the study of families like the Galvins can point to new pathways for treatment that no GWAS can notice. "Somebody once said to me, 'If you genotype every person in the world, will you understand what schizophrenia is?' My guess is you won't understand it just from everybody's genetic sequence. That won't explain schizophrenia. It'll explain a lot about what the risk state represents, but I doubt we're going to have the full answer from that."

DeLisi's work went unnoticed for years. She remains an outsider today—teaching at Harvard Medical School, yes, and active in international schizophrenia research groups, but not recognized with awards or grants like her contemporaries. Even if her SHANK2 findings lead to another breakthrough, she might not get the credit. It's the way of scientific progress—if you aren't among the rare few who are immortalized, you are merely part of the great procession of research, a player in a larger drama. "I think in some ways it bothers me," DeLisi said. "But I have since resolved this in my own mind. It is what I did to make all this possible that counts."

BY THE TIME the SHANK2 study was published, Stefan McDonough had left Amgen. Not long after that, over the phone with her old collaborator, DeLisi learned that McDonough had moved on to Pfizer, the company that had pulled the plug on her multiplex family research sixteen years earlier.

Some small part of her appreciated the irony. If you live long enough, as Mimi Galvin had known, everything comes back to haunt you.

DeLisi had never mentioned any of this to McDonough. As far as he knew, DeLisi's data was all hers; he hadn't known about the big split in 2000, when she got half and Pfizer got half. So, during that call, she decided to let him know that Pfizer still maintained possession of a set of her multiplex family samples, including many of the same families they used for their SHANK2 study.

They both knew what this might mean: Assuming they hadn't been tossed in the trash at some point to make space in a freezer, DeLisi's samples, including the genetic information of the Galvin family, were still sitting somewhere. DeLisi had no idea where—and even if she did, she had no say in how or when or even if it might be used again.

"Who did you deal with here?" McDonough asked. Maybe he could find that person and ask about it.

DeLisi gave him a name.

McDonough couldn't believe it. Thousands of Pfizer employees, all around the world, and the one they were looking for happened, at that very moment, to be sitting just a few feet away from him.

McDonough could hardly resist. It was the end of the year. He had some money left in his budget. "I went ahead and had some of them sequenced," he said. He picked out families with the largest number of relatives with schizophrenia that he could find. The Galvin family had been analyzed already, but there were others, maybe not as big, but big enough.

"Again, Lynn was ahead of her time," McDonough said. "We intend to see if there's anything there. Pfizer won't be interested for its own drug discovery uses, so we have every incentive to publish them and just make the science known to the world."

These families still have something to say. And now someone is listening.

DONALD
JOHN
MICHAEL
RICHARD
MARK
MATTHEW
PETER
MARGARET
LINDSAY

CHAPTER 44

Margaret and Lindsay barely talked or even texted in the six months after Mimi's memorial. The one who cut off contact was Margaret. She saw Lindsay doing so much that it hurt—immersing herself in the Galvin family morass without ever coming up for air, and perhaps even damaging her relationships to her husband and children—and then turning around and admonishing others for not doing the same. Margaret did not see her ever stopping, or even slowing down. "I think there's a lot of manipulation that takes place in our family," Margaret said, "and I think that we've all been on the manipulative side and then the victim side of all of that. And so I find myself as I get older a little bit more assertive with my family, saying, you know, enough is enough."

Only now that their mother wasn't there as a shared focus for them did Margaret see how far apart she and her sister had grown. "Michael and Lindsay don't like it that I don't go in with them on the family dysfunction," Margaret said, "but the boundary is helpful to me."

Lindsay believed that Margaret saying that contact with the family was unhealthy for her was little more than a dodge—an attempt to preempt any criticism that she, Margaret, wasn't helping enough.

As Lindsay saw it, Margaret's passion for self-care was really about her own unresolved fury. "She's got a much higher level of anger towards my mother and my father for how they handled it," Lindsay said. "She has a lot of anger towards my mentally ill brothers, particularly Donald and Jim. I still see a pretty big victim there."

Lindsay repeated something she learned from Louise Silvern, her old therapist, and also from Nancy Gary, and, if she's being honest with herself, from her own mother. "They taught me to embrace the cards you are dealt or it will eat you alive. If you go to the heart of your own matter, you will find only by loving and helping do you have peace from your own trauma." This, in her view, was the major difference between her and her sister.

"We both have worked very hard to save ourselves," Lindsay said. "But she didn't see trying to help them as any part of that, whereas I did."

A few years earlier, Lindsay asked Sam Gary why she wasn't brought to their house like Margaret. "Your parents and I thought you had a stronger constitution," Sam said. "You weren't as fragile." This was news to Lindsay.

But Lindsay was human. She needed help, too. For her entire adult life, when something about the family ate away at her insides, there was only one other person in the world who would understand. When she was at her lowest, her sister was there, living proof that she was not alone. Without Margaret in her life, Lindsay felt as if she'd sustained not one but two losses—a mother and a sister.

"I can't imagine having gone through this without her," Lindsay said.

Hi Gang,

Matt had his vehicle stolen last week after having a new truck totaled a year ago—not his fault and only liability—ugh!

Yeah, right—the poor guy cannot catch a break in life.

Like having schizophrenia is so fun . . .

I just ordered groceries to be delivered to his house tomorrow am. Very easy. https://www.instacart.com . . .

He has no way to go get them and frankly is incapable of grocery shopping.

He would like to move as he is in a really bad area—working

on that with section 8 and the Villanni Family, who said they would have him on one of their buildings. It was fun to see all who knew him at Safeway. "Hey, Matt!"

I would be grateful to anyone of you who could call and offer a hello. No guilt—just asking for some genuine human kindness.

Thanks,

Mary

Email from Lindsay to Margaret, Michael,
John, Richard, and Mark, June 2018

IT WAS UNDERSTOOD among the surviving children of Don and Mimi Galvin that the proceeds from the sale of the house would benefit the three remaining sick brothers. Lindsay brainstormed with Michael about little things that they could do for them with the money. Matt could get a new truck. Peter could get pet therapy or music therapy; even a new tenor recorder might make him happy. Donald loved the opera; what if they hired a companion to take him to those Metropolitan Opera performances they screen at movie theaters?

When she thought about this, Lindsay realized that the person who had really known what her brothers liked, what would make a difference to them, was her mother. This was what kept Lindsay up late now: the idea that the true champion of the family, the gold medal winner in the Empathy Olympics, could have been Mimi Galvin all along. "Now suddenly without her here," Lindsay said, "I'm understanding where she was coming from."

Lindsay used to talk about nature and nurture with her mother. Mimi, still wary of being judged, felt that nurture couldn't have had anything to do with what happened to her family. "Well, it was genetic," she would say. Lindsay told her mother she was not so sure. She believed that some people have a genetic predisposition "that can go either way, depending on your life course and trauma." Certain things can make a difference, Lindsay said, like "love and belonging."

She stopped faulting her mother for this, though. "I really believe that my parents didn't get us as much help as we should have had," she said, "but they didn't know what that looked like."

Lindsay was determined now to channel whatever it was her mother had that helped her connect to the sick boys. So many people—including many of her well brothers—had stopped seeing Donald, Peter, and Matt as human beings a long time ago. Schizophrenia's inaccessibility may be the most destructive thing about it— the thing that keeps so many people from connecting to the people with the illness.

But the mistake—the temptation, especially if you're a relative— is to confuse inaccessibility with a loss of self. "Emotions are always accompanied by some kind of cognitive process," wrote the psychiatrist Silvano Arieti, whose volume *Interpretation of Schizophrenia* dominated the mainstream thinking about the illness in the 1950s and again, with a National Book Award–winning second edition, in the 1970s. "The cognitive process may be unconscious, or automatic, or distorted, but it is always present."

Lindsay noticed this most in her brothers whenever they were on the receiving end of any kindness. "Matt called me this morning with just simple, plain gratitude," she said, shortly after she'd helped him with his groceries. "I wish I could tap into that."

Responding to some gentle prodding, some of her well brothers began reaching out to the sick ones. Richard and Renée called and asked for their phone numbers. Lindsay planned to get Colorado College hockey season tickets for Matt—something Mark might want to take him to, since they once loved playing together. "Pretty much everyone avoids them like the plague. But if I very clearly and deliberately say, 'Hey, can you take them out for, you know, whatever, coffee and a donut?' They'll do it."

IT TOOK SIX months for the sisters to try bridging the gap. They started talking in January, after spending the holidays apart. At the end of a long face-to-face visit, Lindsay started to see things more clearly. "I found myself angry at everybody in my family for not helping me with my mom at the end," Lindsay said. "And Margaret perceived my way of helping as not necessarily a good thing."

Margaret, in turn, acknowledged that Lindsay was more capable of handling the family matters than she ever could have been. But a huge gulf remained between them.

They discussed Margaret's inability to help with Mimi and how angry it made Lindsay. "I just can't do it," Margaret said. And Lindsay felt comfortable enough to say that her sister's decision was not all right with her—that it made her, as she recalled later, "feel sad and frustrated and angry that I feel like I'm left with this whole bag."

They talked a little about survivors of childhood trauma, and how they often continue to find people in their lives to victimize them, so they can continue to get help. Was Lindsay playing that role for Margaret now? Was Margaret for Lindsay?

At the end of the conversation, Lindsay posed a question to her sister: Were they willing to accept each other for who they were? Or were they going to continue down the path of thinking the other person was somehow damaged, and impossible to be close with?

After that visit, Lindsay decided that she needed to allow all of her siblings to do things their way, even as she did things her way. "It's about everyone's own journey," Lindsay said, trying to find some distance of her own. "How they're able to muddle through life and deal."

From her family, Lindsay could see how we all have an amazing ability to shape our own reality, regardless of the facts. We can live our entire lives in a bubble and be quite comfortable. And there can be other realities that we refuse to acknowledge, but are every bit as real as our own. She was not thinking of her sick brothers now, but of everyone—all of them, including her mother, including herself.

"I could just act like I'm a multimillionaire like my brother Richard. Or I could move to Boise like John, or I could play classical guitar all day like Michael. It's, like, we all just *do*. Just respecting that about each other. We all survived somehow. Everyone's different way needs to be okay."

Lindsay was getting closer, finally, to seeing how nature and nurture work together. Her mother had always insisted, defensively, that the illness was genetic, and in a way, Mimi was right. Biology is destiny, to a point; that can't be denied. But Lindsay understood now how we are more than just our genes. We are, in some way, a product of the people who surround us—the people we're forced to grow up with, and the people we choose to be with later.

Our relationships can destroy us, but they can change us, too, and restore us, and without us ever seeing it happen, they define us.

We are human because the people around us make us human.

DONALD

JOHN

MICHAEL

RICHARD

MARK

MATTHEW

PETER

MARGARET

LINDSAY

KATE

JACK

CHAPTER 45

Lindsay's daughter, Kate, grew up to look just like her mother—the same bright eyes, the same relaxed smile. Before having children, Lindsay and Rick, like Margaret and Wylie, had been assured by Dr. Freedman that the chances of passing along mental illness from parent to child—even in the extraordinary case of the Galvin family—were still very small. But parents always worry. And Lindsay had never been one to leave anything to chance.

When she was a little girl, Kate started to flinch and melt down in loud environments like playgrounds and classrooms. These were sensory processing issues. Kate needed occupational therapy, that much was clear. But when you have six mentally ill brothers and your child starts having temper tantrums that you can't control, there is very little to keep you from wondering if this is the beginning of a story that will not end well.

Lindsay thought the worst. She hurled every possible solution she could think of at Kate. She sent her to therapy to learn self-soothing techniques. She bought a hammock for her room, to help her de-stress. She stocked up on essential oils to keep her calm. Was this hyper-vigilance—or just being a proactive, responsible mother? Lindsay didn't know. But something about it worked, or at least it didn't hurt.

Kate thrived. She took all advanced placement classes in her senior year of high school and got straight As in them all—including an art class in which she won an award for a series of works about mental health. Kate got into Berkeley but turned the offer down. Instead, in the fall of 2016, she enrolled at CU Boulder as a sophomore, where she continued to get straight As and spent her summers taking classes. She was, like her mother, a grind—not romantic in the least about childhood, eager to become a grown-up as soon as possible.

In fact, when Kate looked back on her childhood, what she recalled most vividly was how, as soon as she moved past her sensory issues and started doing well, her mother diverted her worry and attention away from her and toward her little brother, Jack.

Jack got therapy as a child, too—prophylactically, just to be on the safe side. He later told his parents that it was all the therapy and testing that made him the most tense. Jack felt put on the spot, like he was being watched all the time. He wasn't wrong: Lindsay and Rick both knew that the Galvin disease affected neither of the girls and six of the boys. Jack was Don and Mimi Galvin's grandson. How could his parents not be watching?

During his freshman year of high school, Jack started skipping class and hanging out in the skateboarding park with a new set of friends. He had been diagnosed with attention deficit disorder, and he'd supplemented his medication with pot. As a teenager, he was engaging in attention-seeking behavior, probably out of boredom; like their mother, Jack and Kate both were academically precocious and had trouble being challenged in a classroom.

For Lindsay and Rick, a pot-smoking male child of the Galvin family was the equivalent of a five-alarm fire. They went searching for people to advise them, and they found just two who understood both the challenges of childhood disorders and the particular issues of their family: Sam and Nancy Gary.

Just after Labor Day in 2015, Jack enrolled in Open Sky, a ninety-day, wilderness-based youth therapy program. One of the most expensive programs of its kind, Open Sky is designed to pull kids out of toxic or dysfunctional environments and reframe their perspective. Its approach is Buddhist, teaching meditation and other techniques to help young people with oppositional disorders and substance issues. The bill was paid by the Garys. "I would not let anything happen to

Mary, or Margaret for that matter," Nancy said. "I would help her do whatever she has to do."

Short programs like Open Sky often serve as a prelude to longer-term treatments. When Jack completed his ninety days, he enrolled in a therapeutic boarding school called Montana Academy. Sam and Nancy paid for that, too—$8,300 a month for twenty-one months. Montana Academy attracts kids with a variety of substance and mental health issues: bulimia, anorexia, anxiety disorders. It was there that Lindsay and Rick learned that Jack's issues had less to do with pot or ADD than with anxiety—the fear of becoming mentally ill.

Jack was angry. He had been saddled with a genetic legacy he'd never asked for, and made to feel like a freak. Lindsay blamed herself for this. "I made such a deliberate effort to expose my children to my mentally ill brothers, so they would not have a bias or feel shame around it. It sort of backfired a little bit."

But it wasn't just the brothers themselves who affected him. For both Jack and his sister, it was witnessing the strain that their mother shouldered, the burden she carried. "My kids have seen how much pain all of it has caused over all the years, and I think they're protective of me," Lindsay said. "Anytime I'm having to deal with something—my sister, or my mom, or one of my brothers—there's angst and frustration around it."

When Lindsay looked at Jack, some part of her had to recognize herself—the little girl she'd once been, walking rings around her brother Donald, tightening the rope, planning to burn him at the stake, bursting with fury and shame.

AT THE START of the program, Lindsay asked her sister to come with her to Montana for moral support as she dropped her son off. The Garys flew them there on their Cessna, just like the old days—always welcoming them, always ready to help. It was a time warp for them both—the meadows in shades of green, yellow, and rust; the dusting of snow on the trees; the gorgeous home; the tennis courts, the orchard, and the horses. Even Trudy, the housekeeper, was still there, embracing both sisters warmly.

That weekend, Margaret's own past replayed in the back of her mind—not just being back in Montana with Sam and Nancy, but watching Lindsay and Rick in the same position her parents must have been in so long ago, when they decided to send her away to the Garys. But she was there to help Lindsay, not to relive the past. Lindsay was going through huge emotional swings. On one hand, Lindsay understood the privileged position she was in. On the other hand, her son was going to be away from her for two whole years. What kind of mother does that? Of course, both she and Margaret knew the answer to that question.

For both sisters, being around the Garys gave them that feeling they had become accustomed to so long ago—the awareness that they were, simultaneously, some of the unluckiest and luckiest people on the face of the earth.

When he got home, Jack did well, attending school, staying sober, and earning good grades again. Jack had learned to manage his anxiety with rock climbing, meditation, even journaling, though he was quick to acknowledge that all these techniques were just deflections. "There's no real way around the anxiety," he said now. "You have to go through it." Jack had become so therapized that he policed everyone else in the house. "He calls us out on our stuff all the time, and uses all the technical language," Lindsay said, awash in relief.

Nancy gave Jack a fly rod as a graduation gift. "He's a different kid," she said. For college, Jack was looking to study early childhood education. After that, he wanted to pursue a career in outdoor wilderness therapy.

When Lindsay looks at Jack now, she thinks not of herself, but of Peter and Donald and Matt and all of her sick brothers. What sort of early interventions might have helped them before the medications took their toll, neutralizing them without curing them? And what about the thousands of people who couldn't afford what her son had—who languish because of a lack of resources, or a stigma from a society that would prefer to pretend that people like them do not exist?

"The haves have these options and the have-nots do not," Lindsay said. "To see this kid take this other track and have it be so successful—it could have easily gone the other direction. I genuinely believe if my brothers had had the opportunity to do something like this, they may not have become as ill as they became."

—◊◊—

In the summer of 2017, at his laboratory in Denver, Robert Freedman took the unusual step of allowing an undergraduate to shadow him in his lab—a young pre-med major from CU Boulder with a special interest in neuroscience. She wanted to be a researcher, like Freedman, focusing on schizophrenia, her family illness.

On a sunny day in June, Kate walked into Freedman's lab for the first time and met the lab techs and assistants, all graduate students, some five years older than she was. When they learned that she was just eighteen, they took notice. This was a highly sought-after position. One of them made a crack about how her family must have been huge donors to get her in there.

Kate smirked. "Well, are you talking money," she asked, "or tissue?"

Lindsay's daughter walked past a room like the one where her mother and aunt and several of her uncles had come to test their auditory gating, listening to those double-clicks with electrodes affixed to their heads, years before she was born. She moved alongside the counters where genetic material from her family and others had been analyzed for evidence of the CHRNA7 irregularity. She stood near where the data from choline trials on little children were studied for signs of schizophrenia—tests that could change everything for a future generation, thanks to six of her uncles.

Her grandfather's brain was probably lying around there someplace. She wondered how long it would take before she could have a look at it.

ACKNOWLEDGMENTS

In early 2016, my great friend Jon Gluck first introduced me to Margaret Galvin Johnson and Lindsay Galvin Rauch. The sisters had been searching for a way to let the world know about their family. They knew that to do their story justice, every living Galvin family member would have to agree to participate—to speak frankly and unreservedly about what, until then, had been private and often very sensitive family issues—and the author would need the independence to follow the story in any direction. I'm extremely grateful that everyone agreed. My deepest thanks to Margaret and Wylie Johnson, Lindsay and Rick Rauch, Peter Galvin, Matthew Galvin, Mark Galvin, Richard and Renée Galvin, Michael Galvin, John and Nancy Galvin, and Donald Galvin—and, most poignantly, Mimi Galvin, who was so willing to open up about her life before her death in 2017. This book is a testament to the entire family's generosity, candor, and faith that their story can be a help to others.

Lindsay and Margaret deserve special acknowledgment. As her mother's executor and the legal authority for her mentally ill brothers, Lindsay worked tirelessly to locate medical records that no one knew still existed, filing reams of paperwork and connecting with a platoon of mental-health professionals and hospital administrators. Margaret, in turn, offered up decades of personal journals and diaries and biographical essays, supplying many priceless details about life on Hidden Valley Road. Both sisters have spent countless hours with me, in person and on the phone and over email, never once balking at the most picayune or intrusive questions or requests. My heartfelt thanks to them both.

I also owe a world of thanks to the psychiatrists and researchers who studied the family—Lynn DeLisi, Robert Freedman, and Stefan McDonough—each of whom spent many hours with me, explaining their research and, with the family's blessing, connecting the dots between their work and the Galvins for the first time publicly. Several other experts in genetics, psychiatry, epidemiology, and the history of science helped me gain a broader understanding of the debates and theories of mental illness: Euan Ashley, Guoping Feng, Elliot Gershon, Steven Hyman, John McGrath, Benjamin Neale, Richard Noll, Edward Shorter, E. Fuller Torrey, and Daniel Weinberger. And I am eternally grateful to Kyla Dunn, whose expertise in genetics helped me ask the right questions in the beginning of this project and saved me from a number of embarrassing errors at the end. (Any errors that remain are, of course, my own.)

My thanks to many additional family members, many of whom are not quoted directly but whose perspectives contributed to the narrative: Eileen Galvin Blocker, Kevin Galvin, Levana Galvin, Melissa Galvin, Patrick Galvin, Betty Hewel, George Hewel, Ellie Johnson, Sally Johnson, Mary Kelley, Kathy Matisoff, Jack Rauch, and Kate Rauch. Thanks also to Nancy Gary (an honorary Galvin if ever there was one), and to the therapists Mary Hartnett and Louise Silvern for their insights into Margaret and Lindsay, and a host of mental-health professionals who have treated the Galvin brothers: Honie B. Crandall, Kriss Prado, Rachel Wilkenson, and, from the Colorado Mental Health Institute at Pueblo, Carmen DiBiaso, Kate Cotner, Sheila Fabrizio-Pantleo, Matthew Goodwin, Julie Meecker, and Al Singleton.

Still others offered insights into specific subjects. Bob Campbell, Jeff Cheney, and Ashley Crockett provided excellent insights into life in Colorado Springs. I'm indebted to many other close family friends and neighbors: Mike Bertsch, Marie Cheney, Ann Crockett, Beck Fisher, Janice Greenhouse, Merri Shoptaugh Hogan, Tim Howard, Ellie Crockett Jeffers, Suzanne King, Ed Ladoceur, Jenna Mahoney, Catherine Skarke McGrady, Roo McKenna, Lynn Murray, Joey Shoptaugh, Carolyn Skarke Solseth, Malham Wakin, and Mark

Wegleitner. For sharing his expertise on falconry, my thanks to Mike Dupuy. For their memories of Don Galvin's falconry heyday, thanks to Jerry Craig, Merrill Eastcott, Relva Lilly, George T. Nolde Jr., Vern Seifert, Hal Webster, and, from the United States Air Force Academy archives, Mary Elizabeth Ruwell. For memories of the Federation of Rocky Mountain States and the Aspen and Santa Fe social scenes, thanks to Nick Jannakos and Robin McKinney Martin. For their unparalleled historical knowledge of the Pueblo mental hospital, thanks to Nell and Bob Mitchell. For their perspectives on Father Robert Freudenstein, thanks to Kent Schnurbusch, Lee Kaspari, Craig Hart, and, from the Catholic chancellery of Denver, Colorado, Douglas Tumminello. For their memories of Brian Galvin, thanks to his former bandmates Scott Philpot, Robert Moorman, and Joel Palmer. And for their memories of Lorelei "Noni" Smith, thanks to Robert Gates, Brandon Gates, and Claudia Shurtz.

For ten years, I've been very lucky to have the support of two extraordinary agents, David Gernert and Chris Parris-Lamb, who believed in this book from the start and led me to the perfect publisher, Doubleday. Thanks to Bill Thomas and Suzanne Herz, and extra gratitude to my editor, the brilliant Kris Puopolo, who, over more than a few plates of taramasalata, helped me understand everything that this book could and should be. Thanks also to Dan Meyer for editorial assistance and photo wrangling, John Fontana for the jacket design, Maria Carella for book design, Rita Madrigal for production management, Fred Chase for copy editing, Dan Novack for his legal review, and Anne Collins of Random House Canada for her extremely helpful read-through of the manuscript. And long before this book was underway, I had racked up debts to many editors who guided me in the past, including Jerry Berkowitz, Robert Blau, Dan Ferrara, Barry Harbaugh, David Hirshey, Adam Moss, Raha Naddaf, Genevieve Smith, and Cyndi Stivers.

In addition to introducing me to the Galvins, Jon Gluck counseled me at every stage of this project. Jennifer Senior helped me reason my way out of countless dead ends and storytelling snarls. They and other friends, colleagues, and loved ones were kind enough to read part or all of this book in earlier stages: Kristin Becker, Kirsten Danis, Kassie Evashevski, Josh Goldfein, Pete Holmberg, Gilbert Honigfeld, Alex Kolker, Caroline Miller, Chris Parris-Lamb, William Reid, and Frank Tipton. Still others helped with enthusiasm, moral support, life-coaching, and guest-room-crashing: Franco Baseggio, Peter Becker, Yvonne Brown, Brewster Brownville, Gabriel Feldberg, Lee Feldshon, Kirsten Fermaglitch, Tony Freitas, David Gandler, Meryl Gordon, Amy Gross, Linda Hervieux, Michael Kelleher, Elaine Kleinbart, Mark Levine, Kevin McCormick, Doug McMullen, Benedict Morelli, Kenneth Mueller, Emily Nussbaum, Saul Raw, Nancy Rome, Phil Serafino, Abigail Snyder, Rebecca Sokolovsky, Clive Thompson, John Trombly, and Shari Zisman. Two researchers, Samia Bouzid and Joshua Ben Rosen, provided great help with selected subjects, and the wonderful Julie Tate performed essential fact-checking.

My mother, Judy Kolker, was my first real reader, the one who before anyone else told me that she could hear my voice when she read my writing. She also spent twenty-five years as a psychiatric counselor at our local hospital in Columbia, Maryland. While reporting this book, I had already started talking about it with her, and I had looked forward to sharing the manuscript with her (and parsing her very careful proofreading). On May 23, 2018—not quite a year after the Galvins lost their matriarch, Mimi—she died at the age of seventy-nine. Her loss has been a blow for our entire family. This book is dedicated to her and to my father, Jon, who, during such a difficult time, has been an incredible model of strength and sensitivity and generosity and grace. I could not have asked for better parents. Much love and gratitude also to my brother, Alex Kolker, and my sister, Fritzi Hallock, both also great role models, and to my entire family, including the Kolkers and Hallocks of Maryland and Iowa and the Danises of Massachusetts, Georgia, and North Carolina.

And finally, to Audrey, whose own writing already shines so brightly, and to Nate, whose advice on structuring the book (and living my life) has spared me a lot of time and heartache. And to my wife, Kirsten, who is so very precious to me—thank you for your love and beauty and inspiration. Everything I write is for you.

Hidden Valley Road is a work of nonfiction drawing from hundreds of hours of interviews with every living member of the Galvin family (including Mimi Galvin, before her death in 2017), as well as dozens of friends, neighbors, teachers, therapists, caregivers, colleagues, relatives, and researchers. No scenes have been invented. All dialogue was either witnessed and recorded by the author or based on published accounts or the recollections of sources who were present at the time.

Additional resources were used to assemble the family narrative—including, most notably, extensive interviews with the schizophrenia researchers Lynn DeLisi, Robert Freedman, and Stefan McDonough; all available medical records for the Galvin brothers and Don Galvin; Don's military service records from the Navy and Air Force; personal correspondence written by Mimi and Don; a series of brief recorded interviews with Mimi, conducted by her daughter Margaret in 2003 and 2008; and several entries from Margaret's personal diaries and autobiographical essays. The text itself makes it clear when any of these sources are being utilized.

For all material requiring further citation—including all passages and chapters about the science of schizophrenia, genetics, and psychopharmacology—notes are provided below.

NOTES

ix   Epigraph: Charles McGrath, "Attention, Please: Anne Tyler Has Something to Say," *New York Times,* July 5, 2018.

CHAPTER 1

4   Marshall Field, Oscar Wilde, and Henry Ward Beecher: Sprague, *Newport in the Rockies.*

5   Don got his hands on a copy: Husam al-Dawlah Timur Mirza, *The Baz-nama-yi Nasiri: A Persian Treatise on Falconry,* trans. Douglas C. Phillott (London: B. Quaritch, 1908).

CHAPTER 2

14   All descriptions of Daniel Paul Schreber's illness are from his memoir, *Memoirs of My Nervous Illness.*

16   King Saul: Freedman, *The Madness Within Us,* 5.

16   Joan of Arc: Ibid.

17   Kraepelin used the term *dementia praecox:* Arieti, *Interpretation of Schizophrenia,* 10.

17   Kraepelin believed that dementia praecox was caused by a "toxin": McAuley, *The Concept of Schizophrenia,* 35, 27.

17   Eugen Bleuler created the term *schizophrenia:* Gottesman and Wolfgram, *Schizophrenia Genesis,* 14–15; DeLisi, *101 Questions & Answers About Schizophrenia: Painful Minds,* xxiii.

18   When Sigmund Freud finally cracked open Schreber's memoir: Bair, *Jung: A Biography,* 149.

18   he had never thought it was worth the trouble to put any of them on the analyst's

couch: Thomas H. McGlashan, "Psychosis as a Disorder of Reduced Cathectic Capacity: Freud's Analysis of the Schreber Case Revisited," *Schizophrenia Bulletin* 35, no. 3 (May 1, 2009): 476–81.

18   "a kind of revelation": *The Freud/Jung Letters,* 214F (October 1, 1910).

18   "director of a mental hospital": Ibid., 187F (April 22, 1910).

18   Freud's *Psycho-Analytic Notes:* Reprinted in Freud, *Complete Psychological Works,* Vol. 12.

18   psychotic delusions were little more than waking dreams: Lothane, *In Defense of Schreber,* 340, cited in Smith, *Muses, Madmen, and Prophets,* 198.

18   All the same symbols and metaphors: *The Freud/Jung Letters,* 214F (October 1, 1910).

18   a fear of castration: Ibid., 218F (October 31, 1910).

18   "Don't forget that Schreber's father was a doctor": Ibid.

18   "uproariously funny" and "brilliantly written": Ibid., 243J (March 19, 1911), cited by Karen Bryce Funt, "From Memoir to Case History: Schreber, Freud and Jung," *Mosaic: A Journal for the Interdisciplinary Study of Literature* 20, no. 4 (1987): 97–115.

19   Jung fundamentally disagreed with him: Karen Funt, "From Memoir to Case History"; and Zvi Lothane, "The Schism Between Freud and Jung over Schreber: Its Implications for Method and Doctrine," *International Forum of Psychoanalysis* 6, no. 2 (1997): 103–15.

19   sparring about this on and off for years: *The Freud/Jung Letters,* 83J (April 18, 1908) and 11F (January 1, 1907).

19   "In my view the concept of libido": Ibid., 282J (November 14, 1911).

19   Jung made that same case again and again: Ibid., 287J (December 11, 1911).

19   "Your technique of treating your pupils": Ibid., 338J (December 18, 1912).

19   "cannot be explained solely by the loss of erotic interest": Jung, *Jung Contra Freud,* 39–40.

19   "He went terribly wrong": Bair, *Jung: A Biography,* 149.

19   schizophrenia affects an estimated one in one hundred people: Most available analyses of the prevalence of schizophrenia drift around this one percent figure. One recent example: Jonna Perälä, Jaana Suvisaari, Samuli I. Saarni, Kimmo Kuoppasalmi, Erkki Isometsä, Sami Pirkola, Timo Partonen, et al., "Lifetime Prevalence of Psychotic and Bipolar I Disorders in a General Population," *Archives of General Psychiatry* 64, no. 1 (January 2007): 19–28.

A more nuanced breakdown of the estimates follows, from Michael J. Owen, Akira Sawa, and Preben B. Mortensen, "Schizophrenia," *Lancet* (London, England) 388, no. 10039 (July 2, 2016): 86–97: "Schizophrenia occurs worldwide, and for decades it was generally thought to have a uniform lifetime morbid risk of 1% across time, geography, and sex. The implication is either that environmental factors are not important in conferring risk or that the relevant exposures are ubiquitous across all populations studied. This view of uniform risk was efficiently dismantled only in 2008 in a series of meta-analyses by McGrath and colleagues [*Epidemiologic Reviews* 30 (2008): 67–76]. They provided central estimates of an incidence per 100,000 population per year of roughly 15 in men and 10 in women, a point prevalence of 4.6 per 1000, and a lifetime morbid risk of around 0.7%. These estimates were based on fairly conservative diagnostic criteria; when broad criteria—including other psychotic disorders such as delusional disorder, brief psychotic disorder, and psychosis not otherwise specified—were applied, the rates were higher by 2–3 times."

20   a third of all the psychiatric hospital beds in the United States: "U.S. Health Official Puts Schizophrenia Costs at $65 Billion." Comments by Richard Wyatt, M.D., chief of neuropsychiatry, National Institute of Mental Health, at a meeting of the American Psychiatric Association. Available online at the Schizophrenia homepage (http://www.schizophrenia.com/news/costs1.html), May 9, 1996.

20   about 40 percent of adults: NIMH statistic, cited in McFarling, Usha Lee, "A Journey Through Schizophrenia from Researcher to Patient and Back," *STAT*, June 14, 2016.

20   One out of every twenty cases of schizophrenia ends in suicide: Kayhee Hor and Mark Taylor, "Suicide and Schizophrenia: A Systematic Review of Rates and Risk Factors," *Journal of Psychopharmacology* (Oxford, England) 24, no. 4, supplement (November 2010): 81–90.

20   Jacques Lacan, the French psychoanalyst: Jacques Lacan, "On a Question Preliminary to Any Possible Treatment of Psychosis," *Ecrits: A Selection,* trans. Alan Sheridan (New York: W. W. Norton, 1977), 200–201, cited by Martin Wallen, "Body Linguistics in Schreber's 'Memoirs' and De Quincey's 'Confessions,'" *Mosaic: A Journal for the Interdisciplinary Study of Literature* 24, no. 2 (1991): 93–108.

20   By the 1970s, Michel Foucault: Foucault, *Discipline and Punish,* 194; and Noam Chomsky and Michel Foucault, *The Chomsky-Foucault Debate,* 33.

20   "Schizophrenia is a disease of theories": Author's interview with Edward Shorter.

CHAPTER 4

32   Frieda Fromm-Reichmann biographical information and Chestnut Lodge historical information, except where specified, is drawn from Fromm-Reichmann, *Psychoanalysis and Psychotherapy,* Foreword by Edith Weigert, v–x.

33   the young man who assaulted Fromm-Reichmann: Fromm-Reichmann, "Remarks on the Philosophy of Mental Disorder" (1946), *Psychoanalysis and Psychotherapy,* 20.

33   the man who kept silent for weeks: John S. Kafka, "Chestnut Lodge and the Psychoanalytic Approach to Psychosis," *Journal of the American Psychoanalytic Association* 59, no. 1 (February 1, 2011): 27–47.

33   the woman who threw stones: Fromm-Reichmann, "Problems of Therapeutic Management in a Psychoanalytic Hospital" (1947), *Psychoanalysis and Psychotherapy,* 147.

33   anyone who said differently might not care enough about the people they were treating: Fromm-Reichmann, "Transference Problems in Schizophrenics" (1939), *Psychoanalysis and Psychotherapy,* 119.

33   the so-called "gas cure": Heinz E. Lehmann and Thomas A. Ban, "The History of the Psychopharmacology of Schizophrenia," *The Canadian Journal of Psychiatry* 42, no. 2 (March 1997): 152–62.

33   Insulin shock therapy: W. C. Shipley and F. Kant, "The Insulin-Shock and Metra-zol Treatments of Schizophrenia, with Emphasis on Psychological Aspects," *Psychological Bulletin* 37, no. 5 (1940): 259–84.

33   Then came the lobotomy: McAuley, *The Concept of Schizophrenia*, 132.

33   Kraepelin . . . turned up little to nothing: Gottesman, *Schizophrenia Genesis*, 82.

33   Ernst Rüdin, became a major figure in the eugenics movement: Martin Brüne, "On Human Self-Domestication, Psychiatry, and Eugenics," *Philosophy, Ethics, and Humanities in Medicine* 2, no. 1 (October 5, 2007): 21.

34   Kallmann called for sterilizing even "nonaffected carriers": Müller-Hill, *Murderous Science*, 11, 31, 42–43, 70.

34   "Every schizophrenic has some dim notion": Fromm-Reichmann, "Transference Problems in Schizophrenics" (1939), *Psychoanalysis and Psychotherapy*, 118.

34   a new vanguard of American psychoanalysts soon embraced: Silvano Arieti, "A Psychotherapeutic Approach to Schizophrenia," in Kemali, Bartholini, and Richter, eds., *Schizophrenia Today*, 245.

34   Joanne Greenberg: Greenberg, *I Never Promised You a Rose Garden*.

34   "There were other powers": Ibid., 83–84.

34   "The sick are all so afraid": Ibid., 46.

35   "Many parents said—even thought": Ibid., 33.

35   "the dangerous influence of the undesirable domineering mother": Fromm-Reichmann, "Notes on the Mother Role in the Family Group" (1940), *Psychoanalysis and Psychotherapy*, 291–92.

35   It was "mainly" this sort of mother: Fromm-Reichmann, "Notes on the Development of Treatment of Schizophrenics by Psychoanalytic Psychotherapy" (1948), *Psychoanalysis and Psychotherapy*, 163–64.

35   "a perversion of the maternal instinct": Rosen, *Direct Analysis*, 97, 101, cited by Carol Eadie Hartwell, "The Schizophrenogenic Mother Concept in American Psychiatry," *Psychiatry* 59, no. 3 (August 1996): 274–97.

36   "American women are very often the leaders": Fromm-Reichmann, "Notes on the Mother Role in the Family Group."

36   "cold," "perfectionistic," "anxious," "overcontrolling," and "restrictive": John Clausen and Melvin Kohn, "Social Relations and Schizophrenia: A Research Report and a Perspective," in Don D. Jackson, *The Etiology of Schizophrenia*, 305.

36   "prototype of the middle class Anglo-Saxon American Woman": Suzanne Reichard and Carl Tillman, "Patterns of Parent-Child Relationships in Schizophrenia," *Psychiatry* 13, no. 2 (May 1950): 253, cited by Hartwell, "The Schizophrenogenic Mother Concept in American Psychiatry."

36   These descriptions seemed to lack a certain coherence: Hartwell, "The Schizophrenogenic Mother Concept in American Psychiatry," 286.

36   the "double-bind": Gregory Bateson, Don D. Jackson, Jay Haley, and John Weakland, "Toward a Theory of Schizophrenia," *Behavioral Science* 1, no. 4 (January 1, 1956): 251–64.

37   "became dangerous figures to males": Lidz, *Schizophrenia and the Family*, 98, 83, cited by Hartwell, "The Schizophrenogenic Mother Concept in American Psychiatry."

CHAPTER 6

52   Geological information about the Woodmen Valley derives from John I. Kitch and Betsy B. Kitch, *Woodmen Valley: Stage Stop to Suburb* (Palmer Lake, Colo.: Filter Press, 1970).

CHAPTER 7

64   "a wastebasket diagnostic classification": McNally, *A Critical History of Schizophrenia*, 153–54.

65   The second edition of the DSM, published in 1968: Seymour S. Kety, ed., "What Is Schizophrenia?," *Schizophrenia Bulletin* 8, no. 4 (1982): 597–600.

CHAPTER 9

75   Except where noted, all material on NIMH's study of the Genain family is from Rosenthal, *The Genain Quadruplets*. Specific citations from that text follow.

75   Every bit as consequential . . . as the case of Daniel Paul Schreber: Irving I. Gottesman, "Theory of Schizophrenia," *The British Medical Journal* 1, no. 5427 (1965): 114.

75   Researchers in Europe and America conducted and published many major twin studies: Mads G. Henriksen, Julie Nordgaard, and Lennart B. Jansson, "Genetics of Schizophrenia: Overview of Methods, Findings and Limitations," *Frontiers in Human Neuroscience* 11 (2017).

75   1928: H. Luxenburger, "Vorläufiger Bericht über psychiatrische Serienuntersuchungen an Zwillingen," *Zeitschrift für die gesamte Neurologie und Psychiatrie* 116 (1928), 297–326.

75   1946: F. J. Kallmann, "The Genetic Theory of Schizophrenia; an Analysis of 691 Schizophrenic Twin Index Families," *American Journal of Psychiatry* 103 (1946), 309–22.

75   1953: Eliot Slater, "Psychotic and Neurotic Illnesses in Twins" (1953), in Slater, *Man, Mind, and Heredity*, 12–124.

76   "When one first learns": Rosenthal, *The Genain Quadruplets*, 7.

76   Nora was the firstborn: Ibid., 362.

76   Iris, meanwhile: Ibid., 16–17.

76   Hester was quiet: Ibid.

76   Myra had a more "sparkling" personality: Ibid., 364.

76   the girls' mother had tried to separate Nora and Myra from Iris and Hester: Ibid., 73.

77   "It is easy to see that": Ibid., 567.

77   the "extreme situation" concept: Ibid., 548.

77   "an atmosphere of fear, suspicion and distrust": Ibid., 566.

77   "We must be more circumspect yet more precise in our theory-building": Ibid., 579.

CHAPTER 10

84   When the hospital first opened with about a dozen patients: Nell Mitchell, *The 13th Street Review*, 7.

84   "We considered it a minor operation": Mike Anton, "Colorado Routinely Steril-
      ized the Mentally Ill Before 1960," *Rocky Mountain News,* November 21, 1999.

84   By the 1950s, the hospital housed more than five thousand patients: Nell Mitchell,
      The 13th Street Review, 47.

85   "These are mostly psychopaths": Telfer, *The Caretakers,* 218.

85   A *New York Times* reviewer called *The Caretakers* a clarion call: Frank G. Slaughter,
      "Life in a Snake-Pit," *New York Times,* November 22, 1959.

85   a scathing thirty-page attack: "Pueblo Grand Jury Blasts State Hospital Program,"
      Colorado Springs Gazette-Telegraph, May 19, 1962.

86   "euphoric quietude": M. Lacomme et al., "Obstetric Analgesia Potentiated by As-
      sociated Intravenous Dolosal with RP 4560," *Bulletin de la Fédération des Sociétés
      de Gynécologie et d'Óbstetrique de Langue Française* 4: (1952): 558–62, cited by
      Bertha K. Madras, "History of the Discovery of the Antipsychotic Dopamine D2
      Receptor: A Basis for the Dopamine Hypothesis of Schizophrenia," *Journal of the
      History of the Neurosciences* 22, no. 1 (January 1, 2013): 62–78.

86   "chemical lobotomy": H. Laborit and P. Huguenard, "L'hibernation artificielle par
      moyens pharmacodynamiques et physiques," *Presse médicale* 59 (1951): 1329, cited
      by Heinz E. Lehmann and Thomas A. Ban, "The History of the Psychopharmacol-
      ogy of Schizophrenia," *The Canadian Journal of Psychiatry* 42, no. 2 (March 1997):
      152–62.

86   side effects: Theocharis Kyziridis, "Notes on the History of Schizophrenia," *Ger-
      man Journal of Psychiatry* 8, no. 3 (2005): 42–48.

87   Arvid Carlsson suggested that Thorazine: Arvid Carlsson and Maria L. Carlsson,
      "A Dopaminergic Deficit Hypothesis of Schizophrenia: The Path to Discovery,"
      Dialogues in Clinical Neuroscience 8, no. 1 (March 2006): 137–42.

87   known as the "dopamine hypothesis": Bertha K. Madras, "History of the Dis-
      covery of the Antipsychotic Dopamine D2 Receptor: A Basis for the Dopamine
      Hypothesis of Schizophrenia."

87   even better than Thorazine: S. Marc Breedlove, Neil V. Watson, and Mark R.
      Rosenzweig, *Biological Psychology,* 5th ed. (Sunderland, Mass.: Sinauer Associates,
      2007), 491.

CHAPTER 13

112   "the existing mediocrity": Sartre, *The Psychology of Imagination,* 169, cited by
      Laing, *The Divided Self,* 84–85.

112   schizophrenia was an act of self-preservation by a wounded soul: Laing, *The Di-
      vided Self,* 73, 75, 77.

112   "lobotomies and tranquilizers": Ibid., 12.

112   a way of playing possum . . . better to turn oneself into a stone: Ibid., 51.

113   sociologist Erving Goffman: McNally, *A Critical History of Schizophrenia,* 149.

113   schizophrenics were almost like prophets: Arieti, *Interpretation of Schizophrenia,*
      125–26.

113   insanity was a concept wielded by the powerful against the disenfranchised: Szasz,
      The Myth of Mental Illness, 188, 176.

113   a war of wits inside of an insane asylum: Kesey, *One Flew Over the Cuckoo's Nest.*

113   "secondary element": Fromm-Reichmann, "On Loneliness" (posthumously pub-
       lished essay), *Psychoanalysis and Psychotherapy,* 328.

114   "If the human race survives": Laing, *The Politics of Experience,* 107.

114   called the family structure a metaphor for authoritarian society: Deleuze and
       Guattari, *Anti-Oedipus,* 34–35.

CHAPTER 14

115   The account of the Puerto Rico conference comes from Rosenthal and Kety, eds.,
       The Transmission of Schizophrenia. Specific citations follow.

116   their study in Denmark: David Rosenthal, "Three Adoption Studies of Heredity
       in the Schizophrenic Disorders," *International Journal of Mental Health* 1, no. 1/2
       (1972): 63–75.

117   a study that reached a very similar conclusion: Irving Gottesman and James
       Shields, "A Polygenic Theory of Schizophrenia," *Proceedings of the National Acad-
       emy of Sciences* 58, no. 1 (July 1, 1967): 199–205.

118   a childhood spent in chaos or poverty could be one cause: Melvin L. Kohn, "Social
       Class and Schizophrenia," in Rosenthal and Kety, eds., *The Transmission of Schizo-
       phrenia,* 156–57.

118   "embittered, aggressive and devoid of natural warmth": Yrjö O. Alanen, "From the
       Mothers of Schizophrenic Patients to Interactional Family Dynamics," in Rosen-
       thal and Kety, eds., *The Transmission of Schizophrenia,* 201, 205.

118   "he perceives very faulty nurturance": Theodore Lidz, "The Family, Language, and
       the Transmission of Schizophrenia," in Rosenthal and Kety, eds., *The Transmission
       of Schizophrenia,* 175.

118   "white-shirted French duel": David Rosenthal, "The Heredity-Environment Issue
       in Schizophrenia: Summary of the Conference and Present Status of Our Knowl-
       edge," in Rosenthal and Kety, eds., *The Transmission of Schizophrenia,* 413.

118   "warring camps": David Reiss, "Competing Hypotheses and Warring Factions:
       Applying Knowledge of Schizophrenia," first presented in 1970 and later published
       in *Schizophrenia Bulletin* 8 (1974): 7–11.

119   "the case for heredity has held up convincingly": Rosenthal, "The Heredity-
       Environment Issue in Schizophrenia," 415.

119   "In the strictest sense, it is not schizophrenia that is inherited": Ibid., 416.

119   "The genes that are implicated": Ibid.

CHAPTER 16

128   a phone call from Noni's boss's wife: "Apparent Murder-Suicide of Lodi Girl, Boy-
       friend," *Lodi News-Sentinel,* September 8, 1973.

CHAPTER 18

145   In 1979, Wyatt's team published research: Daniel Weinberger, E. Fuller Torrey,
       A. N. Neophytides, and R. J. Wyatt, "Lateral Cerebral Ventricular Enlargement
       in Chronic Schizophrenia," *Archives of General Psychiatry* 36, no. 7 (July 1979):
       735–39.

146   "In 1978, Gershon had coauthored": R. O. Rieder and E. S. Gershon, "Genetic

Strategies in Biological Psychiatry," *Archives of General Psychiatry* 35, no. 7 (July 1978): 866–73.

CHAPTER 19

151   one of a handful of pharmacologists tapped by the CIA: "Private Institutions Used in C.I.A. Effort to Control Behavior," *New York Times,* August 2, 1977.

152   "holding tank": Carl C. Pfeiffer, "Psychiatric Hospital vs. Brain Bio Center in Diagnosis of Biochemical Imbalances," *Journal of Orthomolecular Psychiatry* 5, no. 1 (1976): 28–34.

CHAPTER 21

163   "There is a loud telepathic signal here": Gaskin, *Volume One,* 13.

163   six-foot-four: Jim Ricci, "Dream Dies on the Farm," *Chicago Tribune,* October 2, 1986.

163   ex-Marine: "Why We Left the Farm," *Whole Earth Review,* Winter 1985.

163   Monday Night Class: Ibid.

163   OUT TO SAVE THE WORLD: Moretta, *The Hippies,* 232.

163   paid nearly $120,000 for 1,700 acres: Ibid., 232.

164   the nation's largest commune: National Science Foundation estimate, cited by Ricci, "Dream Dies on the Farm."

164   a population of about 1,500 people: Moretta, *The Hippies,* 236.

164   Stephen Gaskin was licensed: Ibid., 233.

164   preferring to marry two couples to one another: Ibid., 240.

164   wholehearted endorsement of tantric sex: Ibid.

164   bountiful supply of homegrown hallucinogenic mushrooms: Ibid., 242.

164   complain that all he had time for all day was settling everyone else's conflicts: Gaskin, *Volume One,* 11, 13, 14.

164   Gaskin controlled: Moretta, *The Hippies,* 238.

165   "thirty dayers": Stiriss, *Voluntary Peasants,* chapter 3, loc. 786, Kindle.

165   "A smart horse runs at the shadow of the whip": Ibid.

165   "six-marriage": Ibid.

165   Four or more babies: Ibid.

165   "a special kind of hippie": Moretta, *The Hippies,* 233.

165   Tibetan yogi Milarepa: Gaskin, *Volume One,* 19–21.

165   "People who live by waterfalls don't hear them": Ibid., 13.

165   the Rock Tumbler: Moretta, *The Hippies,* 240.

166   "constructive feedback" for Farm members who were "on a trip": Stiriss, *Voluntary Peasants,* chapter 3, loc. 218, Kindle.

166   "You are the only variable": Ibid.

CHAPTER 24

179   "vulnerability hypothesis": Joseph Zubin and Bonnie Spring, "Vulnerability—A New View of Schizophrenia," *Journal of Abnormal Psychology* 86, no. 2 (April 1977): 103–26.

179   an update, or elaboration, of Irving Gottesman's 1967 diathesis-stress hypothesis:

Irving Gottesman and James Shields, "A Polygenic Theory of Schizophrenia," *Proceedings of the National Academy of Sciences* 58, no. 1 (July 1, 1967): 199–205.

179   "an opportunity for vulnerability to germinate into disorder": Zubin and Spring, "Vulnerability."

179   "sensory gating": Freedman, *The Madness Within Us,* 35.

179   explanation for the schizophrenia experienced by John Nash: Robert Freedman, "Rethinking Schizophrenia—From the Beginning," Lecture at the Brain and Behavior Research Foundation, October 23, 2015.

179   the "pruning hypothesis": Irwin Feinberg, "Schizophrenia: Caused by a Fault in Programmed Synaptic Elimination During Adolescence?," *Journal of Psychiatric Research* 17, no. 4 (1982–1983): 319–34.

CHAPTER 27

202   "New imaging equipment": Sandy Rovner, "The Split over Schizophrenia," *Washington Post,* July 20, 1984.

203   a review of schizophrenogenic-mother research: Gordon Parker, "Re-Searching the Schizophrenogenic Mother," *The Journal of Nervous and Mental Disease* 170, no. 8 (August 1982): 452–62.

203   a study of the case records of every patient: Anne Harrington, "The Fall of the Schizophrenogenic Mother," *The Lancet* 379, no. 9823 (April 2012): 1292–93.

203   "Frieda . . . embarked on a grand experiment": Ann-Louise Silver, "Chestnut Lodge, Then and Now," *Contemporary Psychoanalysis* 33, no. 2 (April 1, 1997): 227–49.

204   On *The Phil Donahue Show:* Peter Carlson, "Thinking Outside the Box," *Washington Post,* April 9, 2001.

204   "That's the brain disease you are looking at": Modrow, *How to Become a Schizophrenic.*

204   In a study published that same year: Daniel Weinberger and R. J. Wyatt, "Cerebral Ventricular Size: Biological Marker for Subtyping Chronic Schizophrenia," in Earl Usdin and Israel Hanin, eds., *Biological Markers in Psychiatry and Neurology.* New York: Pergamon Press, 1982: 505–12.

204   "Unfortunately there is a segment": Modrow, *How to Become a Schizophrenic.*

204   The latest DSM—the DSM-III: Seymour S. Kety, "What Is Schizophrenia?," *Schizophrenia Bulletin* 8, no. 4 (1982): 597–600.

204   The delusional teenage girl did not have schizophrenia at all: Dava Sobel, "Schizophrenia in Popular Books: A Study Finds Too Much Hope," *New York Times,* February 17, 1981.

208   In 1984, just before meeting the Galvins, he had studied: C. Siegel, M. Waldo, G. Mizner, L. E. Adler, and R. Freedman, "Deficits in Sensory Gating in Schizophrenic Patients and Their Relatives. Evidence Obtained with Auditory Evoked Responses," *Archives of General Psychiatry* 41, no. 6 (June 1984): 607–12.

208   DeLisi used data from her families to confirm: L. E. DeLisi, L. R. Goldin, J. R. Hamovit, M. E. Maxwell, D. Kurtz, and E. S. Gershon, "A Family Study of the Association of Increased Ventricular Size with Schizophrenia," *Archives of General Psychiatry* 43, no. 2 (February 1986): 148–53.

208   testing a possible link between schizophrenia and human leukocyte antigens: Lynn R. Goldin, Lynn E. DeLisi, and Elliot S. Gershon, "Relationship of HLA to Schizophrenia in 10 Nuclear Families," *Psychiatry Research* 20, no. 1 (January 1987): 69–77.

209   The first seemed to confirm: Sarah Henn, Nick Bass, Gail Shields, Timothy J. Crow, and Lynn E. DeLisi, "Affective Illness and Schizophrenia in Families with Multiple Schizophrenic Members: Independent Illnesses or Variant Gene(S)?," *European Neuropsychopharmacology* 5 (January 1995): 31–36.

209   The second failed to find a link between schizophrenia and bipolar illness: Lynn E. DeLisi, Ray Lofthouse, Thomas Lehner, Carla Morganti, Antonio Vita, Gail Shields, Nicholas Bass, Jurg Ott, and Timothy J. Crow, "Failure to Find a Chromosome 18 Pericentric Linkage in Families with Schizophrenia," *American Journal of Medical Genetics* 60, no. 6 (December 18, 1995): 532–34.

209   "I am not a firm believer in environment having an effect at all": Jamie Talan, "Schizophrenia's Secrets: 'Hot Spots' on Chromosomes Fuel Academic, Commercial Studies," *New York Newsday,* October 19, 1999.

209   "It is critical that we avoid premature disillusionment": Kenneth S. Kendler and Scott R. Diehl, "The Genetics of Schizophrenia: A Current, Genetic-Epidemiologic Perspective," *Schizophrenia Bulletin* 19, no. 2 (1993): 261–85.

209   "More than ninety percent of the relatives of schizophrenics": Deborah M. Barnes and Constance Holden, "Biological Issues in Schizophrenia," *Science,* January 23, 1987.

209   The odds of siblings in the same family: Gottesman, *Schizophrenia Genesis,* 102–3.

209   about ten times the chance: Kevin Mitchell, *Innate,* 221.

209   higher, even, than heart disease or diabetes: Ibid.

209–10   The hippocampi of the brains . . . were smaller: R. L. Suddath, G. W. Christison, E. F. Torrey, M. F. Casanova, and D. R. Weinberger, "Anatomical Abnormalities in the Brains of Monozygotic Twins Discordant for Schizophrenia," *The New England Journal of Medicine* 322, no. 12 (March 22, 1990): 789–94.

210   In 1987, Weinberger published a theory: Daniel R. Weinberger, "Implications of Normal Brain Development for the Pathogenesis of Schizophrenia," *Archives of General Psychiatry* 44, no. 7 (July 1, 1987): 660.

211   what he called the "epigenetic landscape": Weinberger and Harrison, *Schizophrenia,* 400.

212   "The risk is passed on": Kevin Mitchell, *Innate, 75.*

CHAPTER 32

246   In 1997, Freedman identified CHRNA7: Robert Freedman, H. Coon, M. Myles-Worsley, A. Orr-Urtreger, A. Olincy, A. Davis, M. Polymeropoulos, et al., "Linkage of a Neurophysiological Deficit in Schizophrenia to a Chromosome 15 Locus," *Proceedings of the National Academy of Sciences of the United States of America* 94, no. 2 (January 21, 1997): 587–92.

246   the first gene ever to be definitively associated with schizophrenia: Carol Kreck, "Mental Institute to Focus on Kids," *Denver Post,* March 3, 1999.

247   By the year 2000, at least five more trouble areas would be isolated: Ann Schrader,

"Schizophrenia Researchers Close in on Genetic Sources," *Denver Post,* August 13, 2000.

247   In 1997, Freedman devised an experiment: Freedman et al., "Linkage of a Neurophysiological Deficit in Schizophrenia to a Chromosome 15 Locus."

247   "important and exciting": Denise Grady, "Brain-Tied Gene Defect May Explain Why Schizophrenics Hear Voices," *New York Times,* January 21, 1997.

248   And when, in 2004, he tested: Laura F. Martin, William R. Kem, and Robert Freedman, "Alpha-7 Nicotinic Receptor Agonists: Potential New Candidates for the Treatment of Schizophrenia," *Psychopharmacology* 174, no. 1 (June 1, 2004): 54–64.

CHAPTER 33

250   In 1994, *The New England Journal of Medicine:* William T. Carpenter and Robert W. Buchanan, "Schizophrenia," *New England Journal of Medicine* 330, no. 10 (March 10, 1994): 681–90.

250   "arguably the worst disease affecting mankind, even AIDS not excepted": "Where Next with Psychiatric Illness?," *Nature* 336, no. 6195 (November 1988): 95–96.

250   "Dr. DeLisi and her collaborators": "Sequana to Participate in Multinational Effort to Uncover the Genetic Basis of Schizophrenia," *Business Wire,* April 20, 1995.

251   "beyond the practical capabilities of a small laboratory": Ibid.

251   the largest single-investigator multiplex family study to date: Ibid.

251   The Human Genome Project: Bijal Trevedi, Michael Le Page, and Peter Aldhous, "The Genome 10 Years On," *New Scientist,* June 19, 2010.

252   In 1995, the cancer researcher Harold Varmus: Samuel H. Barondes, Bruce M. Alberts, Nancy C. Andreasen, Cornelia Bargmann, Francine Benes, Patricia Goldman-Rakic, Irving Gottesman, et al., "Workshop on Schizophrenia," *Proceedings of the National Academy of Sciences of the United States of America* 94, no. 5 (March 4, 1997): 1612–14.

252   Weinberger recalled Zach Hall: Transcript of an interview with Daniel Weinberger, conducted by Stephen Potkin at the 48th annual meeting of the American College of Neuropsychopharmacology in Boca Raton, Florida, December 12, 2007.

253   "thousands of common alleles": Shaun M. Purcell, Naomi R. Wray, Jennifer L. Stone, Peter M. Visscher, Michael C. O'Donovan, Patrick F. Sullivan, and Pamela Sklar, "Common Polygenic Variation Contributes to Risk of Schizophrenia and Bipolar Disorder," *Nature* 460, no. 7256 (August 6, 2009): 748–52.

254   copy number variations (CNVs): James R. Lupski, "Schizophrenia: Incriminating Genomic Evidence," *Nature* 455, no. 7210 (September 2008): 178–79.

254   One GWAS, published in *Nature Genetics* in 2013: Stephan Ripke, Colm O'Dushlaine, Kimberly Chambert, Jennifer L. Moran, Anna K. Kähler, Susanne Akterin, Sarah E. Bergen, et al., "Genome-Wide Association Analysis Identifies 13 New Risk Loci for Schizophrenia," *Nature Genetics* 45, no. 10 (August 25, 2013): 1150–59.

254   Another GWAS, published in *Nature* in 2014: Stephan Ripke, Benjamin M. Neale, Aiden Corvin, James T. R. Walters, Kai-How Farh, Peter A. Holmans, Phil

Lee, et al., "Biological Insights from 108 Schizophrenia-Associated Genetic Loci," *Nature* 511, no. 7510 (July 22, 2014): 421–27.

254   "polygenic risk score": Brien Riley and Robert Williamson, "Sane Genetics for Schizophrenia," *Nature Medicine* 6, no. 3 (March 2000): 253–55. (The complete explanation of the risk score: "Analysis of concordance in first-, second- and third-degree relatives suggests that variants at three or more separate loci are required to confer susceptibility, and that these allelic variants increase risk in a multiplicative rather than additive manner, with the total risk being greater than the sum of the individual risks conferred by each variant.")

254   by about 4 percent: Jonathan Leo, "The Search for Schizophrenia Genes," *Issues in Science and Technology* 32, no. 2 (2016): 68–71.

254   "It's sort of a mindless score": Author's interview with Elliot Gershon.

254   "The guess among my colleagues is that we'll need 250,000 schizophrenia patients": Author's interview with Steven Hyman.

255   "Is it a classical organically based biomedical disorder": Kenneth Kendler, "A Joint History of the Nature of Genetic Variation and the Nature of Schizophrenia," *Molecular Psychiatry* 20, no. 1 (February 2015): 77–83.

256   "a disaster": Joan Arehart-Treichel, "Psychiatric Gene Researchers Urged to Pool Their Samples," *Psychiatric News* (American Psychiatric Association), November 16, 2007.

CHAPTER 34

258   being described as effective, safe, and even relatively painless: Scott O. Lilienfeld and Hal Arkowitz, "The Truth About Shock Therapy: Electroconvulsive Therapy Is a Reasonably Safe Solution for Some Severe Mental Illnesses," *Scientific American,* May 1, 2014.

CHAPTER 35

261   Researchers predisposed against the reflexive use of medication: Whitaker, *Mad in America,* 207–8.

CHAPTER 36

269   Sure enough, with the Galvins, DeLisi and McDonough found something: O. R. Homann, K. Misura, E. Lamas, R. W. Sandrock, P. Nelson, Stefan McDonough, and Lynn E. DeLisi, "Whole-Genome Sequencing in Multiplex Families with Psychoses Reveals Mutations in the SHANK2 and SMARCA1 Genes Segregating with Illness," *Molecular Psychiatry* 21, no. 12 (December 2016): 1690–95.

270   a team from the Broad Institute in Cambridge: Aswin Sekar, Allison R. Bialas, Heather de Rivera, Avery Davis, Timothy R. Hammond, Nolan Kamitaki, et al., "Schizophrenia Risk from Complex Variation of Complement Component 4," *Nature* 530, no. 7589 (February 2016): 177–83.

271   others had conducted separate studies: Audrey Guilmatre, Guillaume Huguet, Richard Delorme, and Thomas Bourgeron, "The Emerging Role of SHANK Genes in Neuropsychiatric Disorders: SHANK Genes in Neuropsychiatric Disorders," *Developmental Neurobiology* 74, no. 2 (February 2014): 113–22. Also see Ahmed

Eltokhi, Gudrun Rappold, and Rolf Sprengel, "Distinct Phenotypes of SHANK2 Mouse Models Reflect Neuropsychiatric Spectrum Disorders of Human Patients with SHANK2 Variants," *Frontiers in Molecular Neuroscience* 11 (2018).

272   "a collection of neurodevelopmental disorders": Thomas R. Insel, "Rethinking Schizophrenia," *Nature* 468, no. 7321 (November 11, 2010): 187–93.

272   Another study of SHANK2 and schizophrenia: S. Peykov, S. Berkel, M. Schoen, K. Weiss, F. Degenhardt, J. Strohmaier, B. Weiss, et al., "Identification and Functional Characterization of Rare *SHANK2* Variants in Schizophrenia," *Molecular Psychiatry* 20, no. 12 (December 2015): 1489–98.

273   The geneticist Kevin Mitchell has noted: Kevin Mitchell, *Innate*, 233–34.

CHAPTER 37

276   Freedman's study about choline was published in 2016: Randal G. Ross, Sharon K. Hunter, M. Camille Hoffman, Lizbeth McCarthy, Betsey M. Chambers, Amanda J. Law, Sherry Leonard, Gary O. Zerbe, and Robert Freedman, "Perinatal Phosphatidylcholine Supplementation and Early Childhood Behavior Problems: Evidence for CHRNA7 Moderation," *The American Journal of Psychiatry* 173, no. 5 (May 2016): 509–16.

276   In 2017, the American Medical Association approved a resolution: Carrie Dennett: "Choline: The Essential but Forgotten Nutrient," *Seattle Times,* November 2, 2017.

CHAPTER 43

319   "Like riding the merry-go-round": Rue L. Cromwell, "Strategies for Studying Schizophrenic Behavior," *Psychopharmacologia* 24, no. 1 (March 1, 1972): 121–46.

320   Hearing Voices Movement: Leudar and Thomas, *Voices of Reason, Voices of Insanity.*

320   many schizophrenia patients experience favorable long-term outcomes without prescription drugs: M. Harrow and T. H. Jobe, "Does Long-Term Treatment of Schizophrenia with Antipsychotic Medications Facilitate Recovery?," *Schizophrenia Bulletin* 39, no. 5 (September 1, 2013): 962–65. Also see M. Harrow, T. H. Jobe, and R. N. Faull, "Does Treatment of Schizophrenia with Antipsychotic Medications Eliminate or Reduce Psychosis? A 20-Year Multi-Follow-up Study," *Psychological Medicine* 44, no. 14 (October 2014): 3007–16.

320   more evidence that psychosis exists on a spectrum: S. Guloksuz and J. van Os, "The Slow Death of the Concept of Schizophrenia and the Painful Birth of the Psychosis Spectrum," *Psychological Medicine* 48, no. 2 (January 2018): 229–44.

321   One meta-analysis, published in 2013: R. J. Linscott and J. van Os. "An Updated and Conservative Systematic Review and Meta-Analysis of Epidemiological Evidence on Psychotic Experiences in Children and Adults: On the Pathway from Proneness to Persistence to Dimensional Expression Across Mental Disorders," *Psychological Medicine* 43, no. 6 (June 2013): 1133–49.

321   another study in 2015; John J. McGrath, Sukanta Saha, Ali Al-Hamzawi, Jordi Alonso, Evelyn J. Bromet, Ronny Bruffaerts, José Miguel Caldas-de-Almeida, et al., "Psychotic Experiences in the General Population: A Cross-National Analysis Based on 31,261 Respondents from 18 Countries," *JAMA Psychiatry* 72, no. 7 (July 1, 2015): 697–705.

321 "early detection and intervention model of care": "Early Detection and Prevention of Psychotic Disorders: Ready for 'Prime Time'?," lecture by Jeffrey Lieberman for the Brain and Behavior Research Foundation, February 12, 2019.

321 so-called "soft interventions": John M. Kane, Delbert G. Robinson, Nina R. Schooler, Kim T. Mueser, David L. Penn, Robert A. Rosenheck, Jean Addington, et al., "Comprehensive Versus Usual Community Care for First-Episode Psychosis: 2-Year Outcomes from the NIMH RAISE Early Treatment Program," *American Journal of Psychiatry* 173, no. 4 (October 20, 2015): 362–72.

321 Australia and Scandinavia: Benedict Carey, "New Approach Advised to Treat Schizophrenia," *New York Times,* December 21, 2017.

322 Lieberman at Columbia is developing: "Early Detection and Prevention of Psychotic Disorders: Ready for 'Prime Time'?," lecture by Jeffrey Lieberman for the Brain and Behavior Research Foundation, February 12, 2019.

322 whether the risk of schizophrenia is linked to the condition of an expectant mother's placenta: Gianluca Ursini, Giovanna Punzi, Qiang Chen, Stefano Marenco, Joshua F. Robinson, Annamaria Porcelli, Emily G. Hamilton, Daniel Weinberger, et al., "Convergence of Placenta Biology and Genetic Risk for Schizophrenia," *Nature Medicine,* May 28, 2018, 1.

323 "half of young school shooters have symptoms of developing schizophrenia": Peter Langman, "Rampage School Shooters: A Typology," *Aggression and Violent Behavior* 14 (2009): 79–86.

323 In 2016, the same year as her SHANK2 study, she published a paper: Lynn E. DeLisi, "A Case for Returning to Multiplex Families for Further Understanding the Heritability of Schizophrenia: A Psychiatrist's Perspective," *Molecular Neuropsychiatry* 2, no. 1 (January 8, 2016): 15–19.

CHAPTER 44

329 "Emotions are always accompanied . . .": Arieti, *Interpretation of Schizophrenia,* 216.

BIBLIOGRAPHY

Arieti, Silvano. *American Handbook of Psychiatry,* Vol. 3. New York: Basic Books, 1959.

————. *Interpretation of Schizophrenia.* 2nd ed., completely revised and expanded. New York: Basic Books, 1974.

Bair, Deirdre. *Jung: A Biography.* Boston: Little, Brown, 2003.

Bentall, Richard P. *Doctoring the Mind: Is Our Current Treatment of Mental Illness Really Any Good?* New York: New York University Press, 2009.

Breedlove, S. Marc, Neil V. Watson, and Mark R. Rosenzweig. *Biological Psychology: An Introduction to Behavioral, Cognitive, and Clinical Neuroscience.* 5th ed. Sunderland, Mass.: Sinauer Associates, 2007.

Brown, Alan S., and Paul H. Patterson, eds. *The Origins of Schizophrenia.* New York: Columbia University Press, 2012.

Buckley, Peter, ed. *Essential Papers on Psychosis.* New York: New York University Press, 1988.

Chomsky, Noam A., and Michel Foucault. *The Chomsky-Foucault Debate: On Human Nature.* New York: New Press, 2006.

Conci, Marco. *Sullivan Revisited—Life and Work: Harry Stack Sullivan's Relevance for Contemporary Psychiatry, Psychotherapy and Psychoanalysis.* Trenton, N.J.: Tangram, 2010.

Cromwell, Rue L., and C. R. Snyder. *Schizophrenia: Origins, Processes, Treatment, and Outcome.* New York: Oxford University Press, 1993.

Davis, Kenneth L., Dennis Charney, Joseph T. Coyle, and Charles Nemeroff, eds. *Neuropsychopharmacology: The Fifth Generation of Progress: An Official Publication of the American College of Neuropsychopharmacology.* Philadelphia: Lippincott Williams & Wilkins, 2002.

Deleuze, Gilles, and Félix Guattari. *Anti-Oedipus: Capitalism and Schizophrenia.* Minneapolis: University of Minnesota Press, 1972.

DeLisi, Lynn E. *100 Questions & Answers About Schizophrenia: Painful Minds*. 2nd ed. Sudbury, Mass.: Jones & Bartlett Publishers, 2011.

Dorman, Daniel. *Dante's Cure: A Journey Out of Madness*. New York: Other Press, 2003.

Eghigian, Greg, ed. *The Routledge History of Madness and Mental Health*. Milton Park, Abingdon, Oxfordshire, and New York: Routledge, 2017.

Foucault, Michel, and Jean Khalfa. *History of Madness*. New York: Routledge, 1961/2006.

Foucault, Michel, and Alan Sheridan. *Discipline and Punish: The Birth of Prison*. London: Penguin, 1975. (References to second Vintage Books ed., 1995.)

Freedman, Robert. *The Madness Within Us: Schizophrenia as a Neuronal Process*. Oxford and New York: Oxford University Press, 2010.

Freud, Sigmund, James Strachey, Anna Freud, and Angela Richards. *The Standard Edition of the Complete Psychological Works of Sigmund Freud*, Vol. 12: *The Case of Schreber, Papers on Technique, and Other Works*. London: Hogarth Press, 1966.

Freud, Sigmund, and C. G. Jung. *The Freud/Jung Letters*. Ed. William McGuire. Trans. Ralph Manheim and R.F.C. Hull. Princeton: Princeton University Press, 1974.

Fromm-Reichmann, Frieda. *Principles of Intensive Psychotherapy*. Chicago: University of Chicago Press, 1971.

———. *Psychoanalysis and Psychotherapy. Selected Papers of Frieda Fromm-Reichmann*. Foreword by Edith Weigert. Chicago: University of Chicago Press, 1974.

Gaskin, Stephen. *Volume One: Sunday Morning Services on the Farm*. Summertown: The Book Publishing Co., 1977.

Gillham, Nicholas W. *Genes, Chromosomes, and Disease: From Simple Traits, to Complex Traits, to Personalized Medicine*. Upper Saddle River, N.J.: FT Press, 2011.

Gottesman, Irving I., and Dorothea L. Wolfgram. *Schizophrenia Genesis: The Origins of Madness*. New York: Freeman, 1991.

Greenberg, Joanne. *I Never Promised You a Rose Garden*. New York: Holt, Rinehart & Winston, 1963.

Hornstein, Gail A. *To Redeem One Person Is to Redeem the World: The Life of Frieda Fromm-Reichmann*. New York: Free Press, 2000.

Jackson, Don D. *The Etiology of Schizophrenia: Genetics, Physiology, Psychology, Sociology*. New York: Basic Books, 1960.

Jaynes, Julian. *The Origin of Consciousness in the Breakdown of the Bicameral Mind*. Boston: Houghton Mifflin, 1976.

Johnstone, Eve C. *Searching for the Causes of Schizophrenia*. Oxford: Oxford University Press, 1994.

Jung, C. G., Sonu Shamdasani, and R.F.C. Hull. *Jung Contra Freud: The 1912 New York Lectures on the Theory of Psychoanalysis*. Princeton: Princeton University Press, 1961.

Kemali, D., G. Bartholini, and Derek Richter, eds. *Schizophrenia Today*. Oxford and New York: Pergamon, 1976.

Kesey, Ken. *One Flew Over the Cuckoo's Nest*. New York: Penguin, 1962.

Laing, R. D. *The Divided Self: An Existential Study in Sanity and Madness*. London: Tavistock, 1959.

———. *The Politics of Experience*. New York: Pantheon, 1967.

———. *Sanity, Madness, and the Family*. London: Penguin, 1964.

Leudar, Ivan, and Philip Thomas. *Voices of Reason, Voices of Insanity: Studies of Verbal Hallucinations*. London and New York: Routledge, 2000.

Lidz, Theodore, Stephen Fleck, and Alice R. Cornelison. *Schizophrenia and the Family*. New York: International Universities Press, 1965.

Lieberman, Jeffrey A., and Ogi Ogas. *Shrinks: The Untold Story of Psychiatry*. 1st ed. New York: Little, Brown, 2015.

Lionells, Marylou, John Fiscalini, Carola Mann, and Donnel B Stern. *Handbook of Interpersonal Psychoanalysis*. New York: Routledge, 2014.

Lothane, Zvi. *In Defense of Schreber: Soul Murder and Psychiatry*. Hillsdale, N.J.: Analytic Press, 1992.

Macdonald, Helen. *Falcon*. London: Reaktion, 2006.

———. *H Is for Hawk*. London: Random House, 2014.

McAuley, W. F. *The Concept of Schizophrenia*. Bristol: John Wright, 1953.

McNally, Kieran. *A Critical History of Schizophrenia*. Basingstoke, UK: Palgrave Macmillan, 2016.

Mitchell, Kevin J. *Innate: How the Wiring of Our Brains Shapes Who We Are*. Princeton and Oxford: Princeton University Press, 2018.

Mitchell, Nell. *The 13th Street Review: A Pictorial History of the Colorado State Hospital (Now CMHIP)*. Pueblo: My Friend, The Printer, Inc., 2009.

Modrow, John. *How to Become a Schizophrenic: The Case Against Biological Psychiatry*. Everett, Wash., and Traverse City, Mich.: Apollyon Press; distributed by Publisher's Distribution Center, 1992.

Morel, Benedict A. *Traite des maladies mentales*. Paris: Masson, 1860.

Moretta, John. *The Hippies: A 1960s History*. Jefferson, N.C.: McFarland, 2017.

Müller-Hill, Benno. *Murderous Science: Elimination by Scientific Selection of Jews, Gypsies, and Others, Germany, 1933–1945*. Woodbury, N.Y.: Cold Spring Harbor Laboratory Press, 1988.

Nasar, Sylvia. *A Beautiful Mind*. New York: Simon & Schuster, 1998.

Niederland, William G. *The Schreber Case: Psychoanalytic Profile of a Paranoid Personality*. Hillsdale, N.J.: Analytic Press, 1984.

Noll, Richard. *American Madness: The Rise and Fall of Dementia Praecox*. Cambridge: Harvard University Press, 2011.

Pastore, Nicholas. *The Nature-Nurture Controversy*. New York: Kings Crown Press, Columbia University, 1949.

Peterson, Roger Tory. *Birds over America*. New York: Dodd, Mead, 1948.

Powers, Ron. *No One Cares About Crazy People: The Chaos and Heartbreak of Mental Health in America*. New York: Hachette, 2017.

Richter, Derek. *Perspectives in Neuropsychiatry; Essays Presented to Professor Frederick Lucien Golla by Past Pupils and Associates*. London: H. K. Lewis, 1950.

Rosen, John N. *Direct Analysis: Selected Papers*. New York: Grune & Stratton, 1953.

Rosenthal, David, ed. *The Genain Quadruplets*. New York: Basic Books, 1963.

Rosenthal, David, and Seymour S. Kety, eds. *The Transmission of Schizophrenia: Proceedings of the Second Research Conference of the Foundations' Fund for Research in Psychiatry, Dorado, Puerto Rico, 26th June to 1 July 1967*. Oxford: Pergamon Press, 1969.

Saks, Elyn R. *The Center Cannot Hold: My Journey Through Madness*. New York: Hachette Books, 2015.

Sartre, Jean-Paul. *The Psychology of Imagination* (1940). London: Routledge, 2016.

Scheper-Hughes, Nancy. *Saints, Scholars, and Schizophrenics: Mental Illness in Rural Ireland*. Berkeley: University of California Press, 1977.

Schiller, Lori, and Amanda Bennett. *The Quiet Room: A Journey Out of the Torment of Madness*. New York: Grand Central Publishing, 2011.

Schreber, Daniel Paul. *Memoirs of My Nervous Illness*. New York: New York Review Books, and London: Bloomsbury, 2001.

Sheehan, Susan. *Is There No Place on Earth for Me?* Boston: Houghton Mifflin, 1982.

Slater, Eliot, James Shields, and Irving I. Gottesman. *Man, Mind, and Heredity: Selected Papers of Eliot Slater on Psychiatry and Genetics*. Baltimore: Johns Hopkins University Press, 1971.

Smith, Daniel B. *Muses, Madmen, and Prophets: Rethinking the History, Science, and Meaning of Auditory Hallucination*. New York: Penguin, 2007.

Sprague, Marshall. *Newport in the Rockies: The Life and Good Times of Colorado Springs*. Athens: Swallow Press/Ohio University Press, 1987.

Stiriss, Melvyn. *Voluntary Peasants: A Psychedelic Journey to the Ultimate Hippie Commune*. Warwick, NY: New Beat Books, 2016. Kindle.

Sullivan, Harry Stack, and Helen Swick Perry. *Schizophrenia as a Human Process*. New York: W. W. Norton, 1974.

Szasz, Thomas. *The Myth of Mental Illness*. New York: Harper & Row, 1961.

Telfer, Dariel. *The Caretakers*. New York: Simon & Schuster, 1959.

Thomas, Philip. *The Dialectics of Schizophrenia*. London and New York: Free Association Books, 1997.

Torrey, E. Fuller. *American Psychosis: How the Federal Government Destroyed the Mental Illness Treatment System*. Oxford: Oxford University Press, 2014.

———. *Schizophrenia and Manic-Depressive Disorder: The Biological Roots of Mental Illness as Revealed by the Landmark Study of Identical Twins*. New York: Basic Books, 1994.

———. *Surviving Schizophrenia: A Family Manual*. New York: Harper & Row, 1983.

Wang, Esmé Weijun. *The Collected Schizophrenias: Essays*. Minneapolis: Graywolf Press, 2019.

Ward, Mary Jane. *The Snake Pit*. New York: Random House, 1946.

Weinberger, Daniel R., and P. J Harrison. *Schizophrenia*. Chichester, West Sussex, and Hoboken, N.J.: Wiley-Blackwell, 2011.

Whitaker, Robert. *Mad in America: Bad Science, Bad Medicine, and the Enduring Mistreatment of the Mentally Ill*. Revised paperback. New York: Basic Books, 2010.

White, T. H. *The Goshawk*. London: Jonathan Cape, 1951.

Williams, Paris. *Rethinking Madness: Towards a Paradigm Shift in Our Understanding and Treatment of Psychosis*. San Francisco: Sky's Edge, 2012.

INDEX

Page numbers in *italics* refer to illustrations.